GREEN MATERIALS AND ENVIRONMENTAL CHEMISTRY

New Production Technologies,
Unique Properties, and Applications

GREEN MATERIALS AND ENVIRONMENTAL CHEMISTRY

New Production Technologies,
Unique Properties, and Applications

Edited by
Abu Zahrim Yaser, PhD
Poonam Khullar, PhD
A. K. Haghi, PhD

Apple Academic Press Inc.
4164 Lakeshore Road
Burlington ON L7L 1A4
Canada

Apple Academic Press, Inc.
1265 Goldenrod Circle NE
Palm Bay, Florida 32905
USA

© 2021 by Apple Academic Press, Inc.

Exclusive worldwide distribution by CRC Press, a member of Taylor & Francis Group

No claim to original U.S. Government works

International Standard Book Number-13: 978-1-77188-861-5 (Hardcover)
International Standard Book Number-13: 978-1-77463-494-3 (Paperback)
International Standard Book Number-13: 978-0-42933-067-4 (eBook)

Library and Archives Canada Cataloguing in Publication

Title: Green materials and environmental chemistry : new production technologies, unique properties, and applications / edited by Abu Zahrim Yaser, PhD, Poonam Khullar, PhD, A.K. Haghi, PhD
Names: Yaser, Abu Zahrim, editor. | Khullar, Poonam, editor. | Haghi, A. K., editor.
Description: Includes bibliographical references and index.
Identifiers: Canadiana (print) 20200219618 | Canadiana (ebook) 20200219839 | ISBN 9781771888615 (hardcover) | ISBN 9781774634943 (softcover) | ISBN 9780429330674 (ebook)
Subjects: LCSH: Sustainable engineering. | LCSH: Sustainable design. | LCSH: Environmental chemistry. | LCSH: Materials—Environmental aspects. | LCSH: Manufacturing processes— Environmental aspects.
Classification: CC TA170 .G74 2020 | DDC 628—dc23

CIP data on file with US Library of Congress

Apple Academic Press also publishes its books in a variety of electronic formats. Some content that appears in print may not be available in electronic format. For information about Apple Academic Press products, visit our website at **www.appleacademicpress.com** and the CRC Press website at **www.crcpress.com**

About the Editors

Abu Zahrim Yaser, PhD

Senior Lecturer, Chemical Engineering Programme, University Malaysia Sabah (UMS), Malaysia, E-mail: zahrim@ums.edu.my

Abu Zahrim Yaser, PhD, is a senior lecturer in the Chemical Engineering Programme at Universiti Malaysia Sabah (UMS). He obtained his PhD from Swansea University (UK), his MSc from the Universiti Kebangsaan Malaysia (Malaysia), and his BE from the University of Malaya (Malaysia). Dr. Zahrim's research mainly focuses on the integrated wastewater treatment and composting and has been leading several projects since 2006. Currently, he is the Deputy Dean at the Faculty of Engineering. Dr. Zahrim is a member of Institution of Chemical Engineers (United Kingdom) and the Board of Engineers, Malaysia.

Poonam Khullar, PhD

Assistant Professor, Department of Chemistry, BBK DAV College for Women, Amritsar, Punjab, India, E-mail: virgo16sep2005@gmail.com

Poonam Khullar, PhD, is Assistant Professor and has total teaching experience of more than twelve years and research experience of more than fourteen years. She is a recipient of many awards and honors, including awards for best teacher and best paper conferred by many prestigious societies of India, such as the Indian Society for Surface Science and Technology, Jadavpur University, Kolkata; and Chemical Research Society of India (Bangalore). She has supervised several students for PhD degrees and four are under supervision. Her specialization includes green chemistry approaches toward the synthesis and applications of metal nanoparticles as drug delivery vehicles. She has one patent and published more than 40 research articles in various reputed international journals published by the American Chemical Society, Royal Society of Chemistry, Elsevier, etc.

A. K. Haghi, PhD

Professor Emeritus of Engineering Sciences, Former Editor-in-Chief, International Journal of Chemoinformatics and Chemical Engineering and Polymers Research Journal; Member, Canadian Research and Development Center of Sciences and Cultures

A. K. Haghi, PhD, is the author and editor of 165 books, as well as 1000 published papers in various journals and conference proceedings. Dr. Haghi has received several grants, consulted for a number of major corporations, and is a frequent speaker to national and international audiences. Since 1983, he served as a professor at several universities. He is former Editor-in-Chief of the *International Journal of Chemoinformatics and Chemical Engineering* and *Polymers Research Journal* and is on the editorial boards of many international journals. He is also a member of the Canadian Research and Development Center of Sciences and Cultures (CRDCSC), Montreal, Quebec, Canada. He holds a BSc in urban and environmental engineering from the University of North Carolina (USA), an MSc in mechanical engineering from North Carolina A&T State University (USA), a DEA in applied mechanics, acoustics, and materials from the Université de Technologie de Compiègne (France), and a PhD in engineering sciences from Université de Franche-Comté (France).

Contents

Contributors

Cristóbal Noé Aguilar
Food Research Department. School of Chemistry, Autonomous University of Coahuila, Saltillo Campus, 25280 Coahuila, México

Raluca Marinica Albu
"Petru Poni" Institute of Macromolecular Chemistry, 41A Grigore Ghica Voda Alley, 700487, Iasi, Romania

Amira Ameer
Faculty of Engineering, University Malaysia Sabah, Jln UMS–88400 Kota Kinabalu, Sabah, Malaysia

Hidayati Asrah
Faculty of Engineering, University Malaysia Sabah, Jln UMS–88400 Kota Kinabalu, Sabah, Malaysia

Andreea Irina Barzic
"Petru Poni" Institute of Macromolecular Chemistry, 41A Grigore Ghica Voda Alley, 700487, Iasi, Romania, E-mail: irina_cosutchi@yahoo.com

A. Yu. Bondar
M.T. Kalashnikov Izhevsk State Technical University, Izhevsk, Russia

Luminita Ioana Buruiana
"Petru Poni" Institute of Macromolecular Chemistry, 41A Grigore Ghica Voda Alley, 700487, Iasi, Romania

Gloria Castellano
Department of Experimental Sciences and Mathematics, Faculty of Veterinary and Experimental Sciences, Valencia Catholic University Saint Vincent Martyr, Guillem de Castro-94, E-46001 Valencia, Spain

Chee-Siang Chong
Faculty of Engineering, University Malaysia Sabah, Jln UMS–88400 Kota Kinabalu, Sabah, Malaysia, E-mail: drchongcs@gmail.com

José D. García-García
Nanobioscience Group. Food Research Department. School of Chemistry, Autonomous University of Coahuila, Saltillo Campus, 25280 Coahuila, México

Nurul Aisyah Abd Hadi
Faculty of Engineering Technology, University Tun Hussein Onn Malaysia, 86400 Parit Raja, Batu Pahat, Johor, Malaysia

Nur Hanis Hayati Hairom
Faculty of Engineering Technology, University Tun Hussein Onn Malaysia, 86400 Parit Raja, Batu Pahat, Johor, Malaysia, E-mail: nhanis@uthm.edu.my

Anna Iliná
Nanobioscience Group. Food Research Department. School of Chemistry, Autonomous University of Coahuila, Saltillo Campus, 25280 Coahuila, México

A. F. Ismail
Advanced Membrane Technology Research Center, University Technology Malaysia, 81310 Skudai, Johor, Malaysia, E-mail: afauzi@utm.my

N. M. Ismail
Chemical Engineering Programme, Faculty of Engineering, University Malaysia Sabah, Jln UMS – 88400 Kota Kinabalu, Sabah, Malaysia, Tel.: +607-5535812, Fax: +607-5581463, E-mail: maizura@ums.edu.my

Nurul Hana Ismail
Faculty of Engineering Technology, University Tun Hussein Onn Malaysia, 86400 Parit Raja, Batu Pahat, Johor, Malaysia

V. I. Kodolov
Basic Research-High Educational Center of Chemical Physics and Mesoscopics, Izhevsk, Russia, M.T. Kalashnikov Izhevsk State Technical University, Izhevsk, Russia

José Luis Martínez-Hernández
Nanobioscience Group. Food Research Department. School of Chemistry, Autonomous University of Coahuila, Saltillo Campus, 25280 Coahuila, México, Tel:(844) 4161238, E-mail: jose-martinez@uadec.edu.mx

Georgina Michelena-Álvarez
Cuban Institute for Research on Sugarcane Derivatives (ICIDCA). Vía Blanca #804, Ciudad de la Habana, Zona Postal 10, Código 11000, Cuba

Abdul Karim Mirasa
Faculty of Engineering, University Malaysia Sabah, Jln UMS–88400 Kota Kinabalu, Sabah, Malaysia, E-mail: akmirasa@ums.edu.my

Abdul Wahab Mohammad
Center for Sustainable Process Technology (CESPRO), Faculty of Engineering and Built Environment, University Kebangsaan Malaysia, 43600 UKM Bangi, Selangor, Malaysia

A. Mustafa
Advanced Membrane Technology Research Center, University Technology Malaysia, 81310 Skudai, Johor, Malaysia

R. V. Mustakimov
Basic Research-High Educational Center of Chemical Physics and Mesoscopics, Izhevsk, Russia, M.T. Kalashnikov Izhevsk State Technical University, Izhevsk, Russia

N. A. H. M. Nordin
Chemical Engineering Department, University Technology PETRONAS, 32610 Tronoh Perak, Malaysia

Sukanchan Palit
43, Judges Bagan, Post-Office-Haridevpur, Kolkata–700082, India, Tel.: 0091-8958728093, E-mails: sukanchan68@gmail.com, sukanchanp@rediffmail.com, sukanchan92@gmail.com

Yu. V. Pershin
Basic Research-High Educational Center of Chemical Physics and Mesoscopics, Izhevsk, Russia

J. Philip
Amal Jyothi College of Engineering, Kanjirappally, Kottayam – 686518, India, E-mail: jp@cusat.ac.in

R. Rakhikrishna
Department of Instrumentation, Cochin University of Science and Technology, Cochin–682 022, India

Stefany Elizabeth Reza-Escandón
Food Research Department. School of Chemistry, Autonomous University of Coahuila,
Saltillo Campus, 25280 Coahuila, México

Raúl Rodríguez-Herrera
Food Research Department. School of Chemistry, Autonomous University of Coahuila,
Saltillo Campus, 25280 Coahuila, México

I. N. Shabanova
Basic Research-High Educational Center of Chemical Physics and Mesoscopics, Izhevsk, Russia,
Udmurt Federal Research Center, Ural Division, Russian Academy of Sciences, Izhevsk, Russia

Dilaelyana Abu Bakar Sidik
Faculty of Engineering Technology, Center of Diploma Studies, University Tun Hussein Onn
Malaysia, 86400 Parit Raja, Batu Pahat, Johor, Malaysia

Jaydip Solanki
Department of Chemistry, Sardar Patel University, Vallabh Vidyanagar, Anand, Gujarat–388120, India

Kiran R. Surati
Department of Chemistry, Sardar Patel University, Vallabh Vidyanagar, Anand, Gujarat–388120, India,
E-mail: kiransurati@yahoo.co.in

R. Eddy Syaizul
Faculty of Engineering, University Malaysia Sabah, Jln UMS–88400 Kota Kinabalu, Sabah, Malaysia

N. S. Terebova
Basic Research-High Educational Center of Chemical Physics and Mesoscopics, Izhevsk, Russia,
Udmurt Federal Research Center, Ural Division, Russian Academy of Sciences, Izhevsk, Russia

Yvonne William Tonduba
Faculty of Engineering, University Malaysia Sabah, Jln UMS–88400 Kota Kinabalu, Sabah, Malaysia

Francisco Torrens
Institute for Molecular Science, University of Valencia, PO Box 22085, E-46071, Valencia, Spain,
E-mail: torrens@uv.es

V. V. Kodolova-Chukhontseva
Basic Research-High Educational Center of Chemical Physics and Mesoscopics, Izhevsk, Russia,
Udmurt Federal Research Center, Ural Division, Russian Academy of Sciences, Izhevsk, Russia

A. Z. Yaser
Chemical Engineering Programme, Faculty of Engineering, University Malaysia Sabah,
Jln UMS–88400 Kota Kinabalu, Sabah, Malaysia

Abbreviations

2D	two-dimensional
3D	three-dimensional
A	answers
AmI	ammonium iodide
APDEMS	(3-aminopropyl)-diethoxymethyl silane
APP	ammonium polyphosphate
APPh	ammonium phosphate
APTES	aminopropyltriethoxysilane
ASs	active sites
ATR	attenuated total reflectance
BN-h	boron nitride
BNNTs	boron nitride nanotubes
BOD	biological oxygen demand
C20	cloisite20
CAT	catalase
CCD	central composite design
CdS	cadmium sulfide
CDs	*cyclo* dextrines
CFC	chlorofluorocarbon
CHOL	cholesterol
CIE	Commission Internationale de l'Eclairage
CMP	condensed matter physics
CNT	carbon nanotubes
CO_2	carbon dioxide
COD	chemical oxygen demand
CPO	crude palm oil
Cs	conclusions
CT	charge transport
Cu/C NC	copper/carbon nanocomposite
CuO	copper oxide
CVD	chemical vapor deposition
DFT	density functional theory
DHA	docosahexaenoic

DNA	deoxyribonucleic acid
DO	dissolved oxygen
DOE	Department of Environment
DSSCs	dye-sensitized solar cells
EA	evolutionary algorithm
EBL	electron blocking layer
EF	electric-field
EL	emissive layer
EPA	eicosapentaenoic
EPP	eco processed pozzolan
EPR	electron paramagnetic resonance
ER	endoplasmic reticulum
ETL	electron transport layer
ETS	emissions trading system
EV	electric-vehicle
FCB	fired clay brick
Fe_2O_3	iron oxide
FESEM	field emission electron microscope
FETs	field-effect transistors
FFB	fresh fruit bunches
FR	free-radical
FRGS	Fundamental Research Grant Scheme
FTIR	Fourier transform infrared spectroscopy
FTO	F-doped tin oxide
GHE	greenhouse effect
GR	graphene
GW	global warming
H	hypotheses
H_2O_2	hydrogen peroxide
H_2SO_4	sulfuric acid
Hb	hemoglobin
HCl	hydrochloric acid
HDL	high-density lipoprotein
HTL	hole transport layer
ICEB	interlocking compressed earth brick
IR	infrared
LAAO	L-amino acid oxidase
LD	low-dimensional

LDL	low-density lipoprotein
LDPE	low-density polyethylene
LEDs	light-emitting diodes
LT	low-temperature
MB	methylene blue
MD	magneto-dielectric
MFM	magnetic force microscopy
$MgSO_4$	magnesium sulfate
MMMs	mixed matrix membranes
MMT	montmorillonite
MO	metal-organic
MOHE	Ministry of Higher Education
NaOH	sodium hydroxide
NCs	nanocomposites
ND	nanodevice
NE	nuclear energy
NFs	nanofilms
Ni/C NC	nickel/carbon nanocomposite
NMP	n-methyl-2-pyrrolidone
NMR	nuclear magnetic resonance
NMs	nanomaterials
NPs	nanoparticles
O_2	oxygen
OER	O_2 evolution reaction
OFET	organic field-effect transistor
OH	hydroxyl
OLEDs	organic light-emitting diodes
OPC	ordinary Portland cement
OPV	organic photovoltaics
OSs	oxidation states
Pa	paradoxes
PANI	polyaniline
PAs	pyrrolizidine alkaloids
PDMS	polydimethylsiloxane
PEDOT	poly(3,4-ethylene dioxythiophene)
PEG	polyethylene glycol
PEPA	polyethylene polyamine
PES	polyethersulfone

pHEMA	poly(2-hydroxyethyl methacrylate)
PHOLEDs	phosphorescent organic light-emitting diodes
PL	periodic law
PLA	polylactic acid
PLGA	perylene, polylactic-co-glycolic acid
POFA	palm oil fuel ash
POMs	polyoxometallates
POMSE	palm oil mill secondary effluent
PPF	pillared porphyrin framework
PPy	polypyrrole
PSS	poly(styrene sulfonate)
PTE	periodic table of the elements
PTFE	poly(tetrafluoroethylene)
PU	polyurethane
PV	photovoltaic
PVA	polyvinyl alcohol
Q	questions
QSPRs	quantitative structure-property relationships
RA	risk assessment
RBD	refined, bleached, and deodorized
REs	renewable energies
RH	relative humidity
RNA	ribonucleic acid
RSM	response surface methodology
SAXS	small-angle x-ray scattering
SBE	spent bleaching earth
SBEO	spent bleaching oil
SC	superconductivity
SDE	simultaneous distillation-extraction
SDS	sodium dodecyl sulfate
SEM	scanning electron microscope
SI	selectivity index
SnO_2	tin dioxide
SO_2	sulfone groups
SOD	superoxide dismutase
SPE	solid-phase extraction
STLDs	STL derivatives
STLs	sesquiterpene lactones

SWNTs	single-wall carbon nanotubes
TADF	thermally activated delayed fluorescence
TAPC	1,1-bis[(di-4-tolylamino)phenyl]cyclohexane
TCTA	tris(4-carbazoyl-9-ylphenyl)amine
T_d	degradation temperature
TEM	transmission electron microscopy
TF	thin film
TFC	twice functionalized clay
TFTs	thin-film transistors
T_g	transition temperature
TiO_2	titanium dioxide
TMDs	transition metal dichalcogenides
UPLC	ultra-high-performance liquid chromatography
UV	ultraviolet
VDW	van der Waals
VOCs	volatile organic compounds
VSM	vibrating sample magnetometer
WF	Wiedemann-Franz
WHO	World Health Organization
WL	wavelength
WW2	World War II
XRD	x-ray diffraction
ZIF	zeolitic imidazole framework
ZnO	zinc oxide

Preface

The world faces significant challenges as population and consumption continue to grow while nonrenewable fossil fuels and other raw materials are depleted at ever-increasing rates. Moreover, environmental consciousness and a penchant for thinking in terms of material cycles have caught on with consumers: the use of environmentally compatible materials and production methods is desired, even taken for granted by the client.

Green Materials and Environmental Chemistry is a technical approach to address these issues using green design and analysis. This book provides an overview of the latest developments in environmental chemistry and sustainable materials written by experts in their respective research domains.

The scope of green materials and environmental chemistry depends upon one's perspective and discipline, but it is broadly defined as minimizing environmental impacts across all life cycle phases in the design and engineering of products, processes, and systems. It is important to realize that green materials and environmental chemistry is only one possible approach to addressing the larger issue of sustainability that includes environmental, economic, and social aspects. Green materials and environmental chemistry is necessarily interdisciplinary and, therefore, is best considered as a set of concepts that can be applied across engineering disciplines. Sustainability is the order of the day and the magic word for a better future in politics and industry.

This volume bridges the gap between research and industry on the one hand and designers on the other by offering a systematic overview of the currently available sustainable materials and providing the reader with all the information he or she needs to assess a new material's suitability and potential for a given project. Along the way, it examines natural and biodegradable materials, while also presenting materials with multifunctional properties.

CHAPTER 1

Mixed Matrix Membranes for CO_2/CH_4 Separation with Natural Clay Cloisite15A® Modified with 3-Aminopropyltryethoxysilane as a Coupling Agent

N. M. ISMAIL,[1] A. F. ISMAIL,[2] A. MUSTAFA,[2] A. Z. YASER,[1] and N. A. H. M. NORDIN[3]

[1]*Chemical Engineering Programme, Faculty of Engineering, University Malaysia Sabah, Jln UMS–88400 Kota Kinabalu, Sabah, Malaysia, Tel.: +607-5535812, Fax: +607-5581463, E-mail: maizura@ums.edu.my (N. M. Ismail)*

[2]*Advanced Membrane Technology Research Center, University Technology Malaysia, 81310 Skudai, Johor, Malaysia, E-mail: afauzi@utm.my (A. F. Ismail)*

[3]*Chemical Engineering Department, University Technology PETRONAS, 32610 Tronoh Perak, Malaysia*

ABSTRACT

In this chapter, twice-functionalized organoclay (TFC) was prepared by modifying Cloisite15A with 3-aminopropyltriethoxysilane (APTES). The role of the silane coupling agent in improving the morphology, thermal, mechanical, and gas separation properties of asymmetric flat sheet mixed matrix membranes (MMMs) was investigated. 0.25wt% of modified C15A was embedded in polyethersulfone (PES) matrix (PES/C15AS0.25) and the properties were compared with neat PES and unmodified clay-PES MMM (PES/C15A0.25). Results from FTIR-ATR spectras, XRD diffractograms, SEM, and FESEM micrographs showed characteristics of a partially modified surface of C15A with aminosilane. A significant increase in nitrogen content in PES/C15AS0.25 shown by elemental composition analysis by EDX was in agreement with the potential of aminosilane bonding. Furthermore, the thermal analysis also revealed a slight reduction on the onset of degradation temperature at 145°C. Residual mass content for PES/C15AS0.25 was also 12% and 26% lower than PES and PES/C15A0.25, respectively which confirms the presence of APTES. In addition, not only was the elongation at break of PES/C15AS0.25 was enhanced by TFC embodiment but so was the tensile strength. For CO_2/CH_4 separation, a significant reduction in permeance of both gases for PES/C15AS0.25 showed that the interfacial voidage was reduced by the aminosilane coupling agent, consequently improving the selectivity. In this work, the occurrence of interfacial interaction between the epoxy groups from the silylated C15A and PES matrix attributed to the enhanced morphological, mechanical, and gas separation properties of the PES/C15AS0.25 compared to neat PES and PES/C15A0.25.

1.1   INTRODUCTION

In the last decades, the application of layered silicates as green material additives for polymers has gained significant attention due to their high aspect ratio, layered morphology, intercalative, and exfoliative characteristics, abundantly cheap and easily available [1–3]. In addition, highly dispersed silicate layers allow remarkable improvement on the mechanical, thermal, and barrier properties of the nanocomposites (NCs) [4–11]. However, the incorporation of clay particles in the polymer matrix is challenging due to the poor interfacial interaction between the clay hydrophilic reaction sites and the hydrophobic polymer chains. Therefore, the clay surface often undergoes chemical modification to improve the compatibility with the polymer matrix [12]. Surfactant modification through cation exchange reactions with a quaternary ammonium salt and silylation reaction by organo alkoxysilanes are the two most common methods investigated for clay surface modification [13, 14].

Clay minerals, montmorillonite (MMT) for example, contains active sites (ASs) known as silanols and separated by exchangeable Na^+ and/or Ca^{2+} interlayer cationsis suitable for chemical modification. For MMT, cation exchange occurs when the quaternary ammonium salt or surfactants enters the silicate layer galleries and replaces the available Na^+ and/or Ca^{2+}. Organic groups attached on the layered surface increase the interlayer distance, *d*-spacing. Mechanism of improvements works by the higher penetration of polymer chain into the galleries due to the expansion of *d*-spacing. This in turn increases the polymer-clay interaction and clay dispersion. To date, several works have reported successful surface modification of clay by using surfactant [14, 15]. Despite the increase in the interlayer distance, the interfacial linkage of between the clay surface and polymer matrix is still inefficient [16]. Meanwhile, the presence of hydrophilic edges organoclays will also hinder the intercalation and leads to agglomeration of clay particles in the bulk polymer [13]. A lot of efforts have been carried out to functionalize the clay surfaces effectively by a novel silylation method, or silane grafting to solve the interfacial problem [17, 18]. Several types of alkoxysilanes have been proposed to modify the silicate layers from the MMT family [19, 20]. Organosilanes usually consist of a short hydrocarbon chain with a reactive functional group (X) namely amine, epoxy, or isocyanate and a hydrolyzable group (RO) such as methoxy, ethoxy, or acetoxyin the general formula of $(RO)_3SiCH_2CH_2CH_2$-X. This approach is considerably preferred since after the clay was modified with cation exchanges, the silanol groups remained intact at the silicate layers. Therefore, using silylation method, the epoxy-functional groups from the silane will react with the silanol groups to form the anchoring effect. Since the interlayer space is already occupied by the surfactant, the sylilation reaction takes up the hydroxyl group at the platelet edges due to the presence of broken bonds. Consequently, a rigid network through covalent bonds between the clay mineral, polymer matrix, and silane contributes to the superior properties of the NCs.

Further modification of surfactant modified clay, also known as twice functionalized clay (TFC) showed several advantages and has attracted great interest. For example, Joo, and co-workers [21] observed an increase in tensile modulus for polypropylene/clay nanocomposite incorporating Cloisite20 (C20A) modified with three types of silanes; propyltrimethoxysilane, octytrimethoxysilane, and octadecyltrimethoxysilane compared to unmodified clay. Most notably, the tensile modulus of nanocomposite prepared from modified C20A with octadecyltrimethoxysilane was 1.5 times higher than unmodified clay. Chen et al. [22] prepared poly (L-lactide) (PLLA)/clay with TFC from Cloisite25A (C25A) modified with (glycidoxypropyl)trymethoxysilane (GPS). They demonstrated that the tensile modulus and elongation at break of PLLA/TFC at low loadings were better than nanocomposite of unmodified C25A (PLLA/C25A) and neat PLLA due to higher degree of exfoliation and enhanced interfacial interaction.

Similarly, like other MMMs, poor interfacial adhesion due to clay-polymer incompatibility often leads to the deterioration of membrane performances. Therefore, the aim of this study was to improve the bonding properties of polymer-clay through clay silylation. Besides, most studies are concentrating on the mechanical enhancement from TFC and the reports on the effect on performance for gas separation are rare. For these reasons, this study is carried out to investigate the influence of silylation on the morphological, thermal, mechanical, and the overall effects on the gas separation performance of polyethersulfone (PES) mixed matrix membranes (MMMs). In this study, C15A which was initially modified with quaternary ammonium salt and dimethyl dihydrogenated tallow was further modified by using three aminopropyl triethoxysilane (APTES). MMMs embedded with modified C15A was compared with unmodified C15A and neat PES and subsequently characterized by means of XRD, FTIR, SEM, FESEM-EDX, TGA, DSC, tensile test, and gas permeation measurement. A moderate environment during the modification process such as low temperature and without acidification treatment was attempted to study the overall effect on the resultant MMM.

1.2   EXPERIMENTAL

1.2.1   MATERIALS

PES, under the trade name PES, RADEL A-300A obtained from Solvay Advanced Polymers (Alpharetta, GA, USA) with an average molecular weight of 15 000 g/mol was used for the continuous phase. Commercial cloisite15A, organophilic clay modified with quaternary ammonium salt and dimethyl dihydrogenated tallow (2M2HT), was purchased from Southern Clay Products, Inc., USA. The surface of the clay was modified with 3-aminopropyltriethoxysilane (APTES) (Aldrich Chemical Company, USA) (99%, ρ = 0.946 g/mL). Anhydrous n-methyl-2-pyrrolidone (NMP) and ethanol received from Merck were used as solvent and non-solvent for the membrane fabrication. PES and C15A underwent a heat pre-treatment in the oven at 60°C to remove the moisture content prior to membrane preparation. Polydimethylsiloxane (PDMS) (Sylgard-184) obtained from Dow Corning Corp. was used to coat the membranes. Unless otherwise stated, all chemicals were of analytical grade and used as received.

1.2.2   CLAY SURFACE MODIFICATION

In the present study, the grafting method from Shen et al. [23] was adapted to modify the surface of the clay. A 2.5 g of C15A particles were dispersed in a mixture of ethanol/water (75/25 by volume). A 2 g of APTES was added dropwise into the previous mixture and was continuously sheared for 8 h at room temperature. The modified clay was washed with the same ratio of ethanol/water mixture to remove unreacted silane. Finally, the modified clay was filtered and dried at 60°C in the oven for the removal of trapped water/ethanol mixture. Figure 1.1 shows the steps in the modification of C15A in the ethanol-water mixture medium.

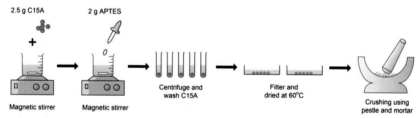

FIGURE 1.1   Aminosilane-clay modification procedure.

1.2.3   MEMBRANE FABRICATION

To fabricate MMM, 0.25wt% of C15A was dispersed in 49.13 g of NMP in a 250 ml flask. 10wt% from the total weight of the polymer was added into this solution and stirred for 4 hours for priming. The remaining polymer was added gradually and stirred until complete dissolution of the polymer. Ethanol was added gradually and the solution was mixed until a homogenous solution was achieved. Prior to casting, the solution was sonicated to remove trapped air bubbles and kept overnight. Then, a small amount of the solution was spread onto a clean, flat glass plate. The thin film (TF) was immersed in a non-solvent water bath at room temperature for overnight. Finally, membranes were air-dried for 3 d at room temperature. Similar membrane fabrication protocol as depicted in Figure 1.2 was applied for the modified C15A while the preparation of the neat membrane was similar to MMM, with the exception of clay addition. Finally, the membranes were coated with 3wt% of PDMS coating solution to fill the pinholes and surface defects prior to gas permeation measurement. MMMs fabricated with modified C15A, or TFC is denoted as PES/C15AS0.25 while PES/C15A0.25 is referred to as unmodified C15A MMM.

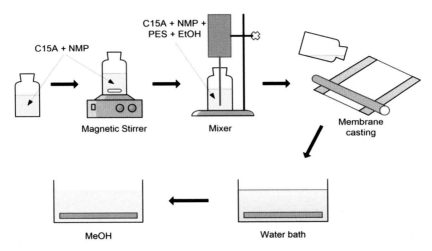

FIGURE 1.2   Membrane fabrication protocol.

1.2.4   CLAY AND MEMBRANE CHARACTERIZATION

The morphology of the clay particles, membranes surface and cross-sectional were investigated by scanning electron microscope (SEM) (TM3000, Hitachi). The cross-sectional morphology of the MMMs was further examined by field emission electron microscope (FESEM) (JEOL JSM-7600F) for higher magnifications. Membrane films were freeze-fractured in liquid nitrogen to reveal the cross-sectional area. These membranes were sputter-coated with gold and platinum mixture prior to observation.

Fourier transform infrared spectroscopy (FTIR) (Nicolet 5700, Thermo Electron Scientific Instruments Corporation, USA) was used to confirm the successful modification of clay and membranes by APTES. 16 scans were collected for each sample within the frequency range of 4000–400 cm^{-1}.

The thermal decomposition profile of the membranes was measured using Thermogravimetry (TGA, Mettler Toledo TGA/SDTA851). Samples were heated from 50–1000°C at 10°C/min under a nitrogen flow. DSC analysis was performed by using Mettler Toledo DSC822e to obtain the glass transition temperature (T_g) of the membranes. The membranes first heating cycle was from 25°C to 250°C with the heating and cooling rate of 10°C/min. A similar heating cycle was repeated to determine the glass transition temperature (T_g).

X-ray diffraction (XRD) patterns were obtained by using Siemens D5000 diffractometer with a scanning range of 2θ was between 2° and 40° with copper Kα ($\lambda = 0.1542$ nm). d-spacings were determined by Bragg's equation:

$$d = \frac{n\lambda}{2sin\theta} \tag{1}$$

where, n is 1 (order of diffraction in the calculation), λ is 0.154056 nm (wavelength (WL) of X-ray), and θ as the angle of the maximum peak in the XRD diffractograms.

Elongation at break and tensile strength of the membranes were obtained by the Universal testing machine (Instron 5567). ASTM D638 (Standard Test Methods for Tensile Properties) was employed during testing. The samples were prepared according to the standard shape and tested for a crosshead speed of 10 mm/min.

Pure gas permeation of CO_2 and CH_4 were performed by a constant volume variable-pressure method. Selectivities and permeances of the gases were evaluated using gas permeation set-up described elsewhere [24]. Membranes with an effective permeation area of 13.5 cm^2 were placed on sintered metal plate and pressurized at the inlet. In order to achieve a steady-state for each measurement, the membranes were exposed to pure gas flow at very low pressure for 10 min. Repeated pure gas permeation experiments were conducted at room temperature and 3 bars. Pressure normalized flux, P/l or permeance (in GPU), was calculated by using the following equation:

$$\frac{P_i}{l} = \frac{1}{A\Delta p}\frac{dV_i}{dt} \tag{2}$$

where, i represents the respective gas penetrant i, $\frac{dV_i}{dt}$ is the volumetric flowrate of gas permeated through the membrane (cm^3/s), A is the effective membrane area (cm^2), and Δp is the pressure drop difference across the membrane (cmHg). Permeance or pressure-normalized gas flux is commonly represented by GPU (1 GPU = 1 x 10^{-6} cm^3 cm/cm^2 s cm Hg). The ideal gas selectivity, α for gas i to gas j is determined by the ratio of permeance of gas penetrant i and j, P_i/l and P_j/l, respectively as below:

$$\alpha_i = \frac{(P_i / l)}{(P_j / l)} \tag{3}$$

1.3   RESULTS AND DISCUSSION

1.3.1   *FOURIER TRANSFORM INFRARED SPECTROSCOPY-ATTENUATED TOTAL REFLECTANCE (FTIR-ATR) ANALYSIS*

FTIR-ATR was used in order to verify the successful surface modification of clay using aminosilane and to identify the respective organic moieties on the clay platelets. Figure 1.3 shows the FTIR spectra for the unmodified and modified Cloisite15A and both revealed a typical MMT absorption bands. The MMT IR spectra shows the presence of strong absorption band at ~1020 cm^{-1} corresponding to the Si-O-Si stretching vibrations and 918 cm^{-1} from Al-OH-Al (aluminates deformation) [16, 25]. In addition, broad band's 3000–3600 cm^{-1} and 1655 cm^{-1} corresponds to the stretching and bending vibrations of adsorbed water were also observed [25].

For unmodified C15A, three absorbance peaks observed at 2850, 2920 and 1468 cm^{-1} represents the intercalation by surfactant (symmetric and asymmetric stretching and bending of C-H vibrations, respectively) [25, 26]. Interestingly, similar absorbance peaks have been reported for the silylated C15A which corresponds to the symmetric and asymmetric stretching vibrations of -CH$_2$-of the alkyl chains for silane moieties, respectively, but with lesser intensity [16, 18, 23]. The removal of excess organic modifier during washing had possibly reduced the methylene groups. This explains the marginal decrease in the peak intensity after aminosilane modification as shown in Figure 1.3 [27]. Kim et al. [20] observed similar results after Cloisite25A was modified with APTES.

Besides the C-H groups, other evidence of successful modification of aminosilanes on the clay surface is the presence of N-H absorbance peaks (~3370 and ~3290 cm^{-1}). However, the respective absorbance peaks were not observed. Since the intensity of this shoulder peak was not apparent even at higher concentration [16, 18], the disappearance of this peak in the present work is expected. Overlapping between the epoxy group and Si-O-Si peaks slightly complicates the analysis, but the higher intensity observed for modified C15A suggested the presence of silane moieties [28].

Further analysis on both spectras exhibited a weak band at ~3629 cm^{-1} and this corresponds to the Si-OH stretching vibration in MMT [16, 18]. For modified C15A, the stretching of this peak should be less intense than that of pristine C15A to confirm the grafting of silanol groups with hydrolyzed silane monomers or oligomers. However, as seen in Figure 1.3, the peak intensity before and after aminosilane modification is identical, which contradict with previous studies [20, 22]. This shows the possibility of very minimal chemical reaction interaction between the silanol groups of C15A and APTES monomers or oligomers which were not reflected in the FTIR-ATR spectra. These results suggested that the moderate modification environment and parameters (without acid treatment and low temperature) used in this work, may attribute to the low hydrolyzation of the APTES hence the difficulty to achieve good silylation [13, 29, 30]. However, further analyses are required to verify the extent of C15A modification and the results will be discussed in the subsequent sections.

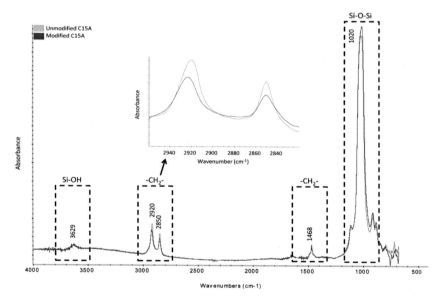

FIGURE 1.3   FTIR-ATR spectra for unmodified and modified Cloisite15A.

To further elaborate the modification process, the proposed mechanism for the modification path of surfactant and aminosilane modified is shown in Figure 1.4. After hydrolysis, alkoxy groups (RO: methoxy, ethoxy, or acetoxy) from APTES were converted to hydroxyl groups. These groups reacted with the hydroxyl groups on the edge of the clay to form silylated clay. Functional groups (amino) from the APTES will further react with the PES polymer chains to form stronger chemical bonds.

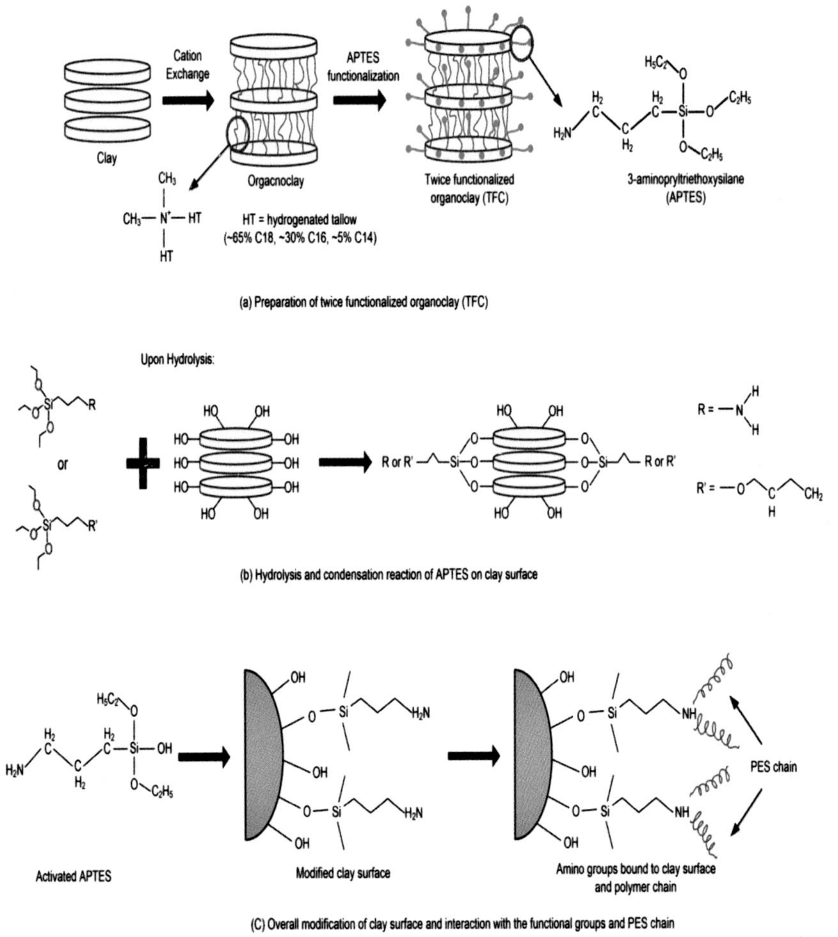

FIGURE 1.4 Proposed mechanism for aminosilane modification on C15A and its interaction with PES polymer chains.

FTIR spectroscopy for PES, PES/C15A0.25, and PES/C15AS0.25 shown in Figure 1.5 was used to identify the various interactions between different components in the membrane at the molecular level. From Figure 1.5, the infrared (IR) bands at 1106 cm⁻¹ and 1150 cm⁻¹ correspond to the symmetric stretching vibration for sulfone groups (SO_2) of PES. The absorption bands displayed at 1298 and 1322 cm⁻¹ are attributed to the SO_2 asymmetric stretch. In addition, another band was also observed at 1240 cm⁻¹ showing the presence of aromatic ether (-C-O-C-) linkage. Meanwhile, the aromatic benzene ring was shown by the characteristics bands at 1486 cm⁻¹ and 1578 cm⁻¹. These observations are comparable with those reported in the literatures [31, 32]. As seen in Figure 1.5, the FTIR spectra of MMMs are similar to those obtained for neat PES. Additionally, one can also observe that all MMMs did not display any characteristic absorption bands of methylene group for C15A (~2850 cm⁻¹ and ~2920 cm⁻¹) [25]. Furthermore, Mokhtar et al. [26] reported that incorporation of C15A revealed another weak absorption band at ~1740 cm⁻¹ corresponding to the stretching vibration of the carbonyl group (=C=O) for MMT [26], which is different from our results. Their work also showed a minimal increase in intensity for these bands even at clay loading up to 10wt%. Hence, for this work, we presumed that the disappearance of these three important absorption bands can be correlated with the use of a relatively very low C15A loading.

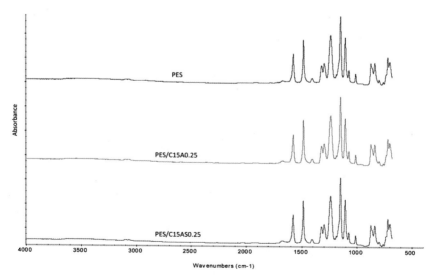

FIGURE 1.5   FTIR-ATR spectra for neat PES and MMMs.

1.3.2 X-RAY DIFFRACTION (XRD) ANALYSIS

The grafting of APTES onto C15A was studied by XRD. In Figure 1.6, the XRD patterns of unmodified and modified C15A are compared. Generally, typical MMT clay are made of several hundreds of individual platy particles (1 nm in thickness, 50–100 nm in lateral dimension) and held by a weak electrostatic force [33]. In addition, it is also known that the introduction of organic modifier into MMT clay through ion exchange often leads to higher d-spacing, normally over 2 nm [34]. As seen in Figure 1.1, the maximum diffraction pattern of the unmodified C15A is located at $2\theta =$ 2.75° which is equivalent to 3.21 nm. The higher d-spacing from those reported for typical MMT indicates the presence of salt molecules into the C15A structure [6, 35]. In addition, this analysis also revealed two natural MMT peaks at 2 of 7.2° and 19.75°. The peaks correspond to d-spacing of 1.23 and 0.45 nm for planes d_{001} and d_{002}. A comparable diffractogram pattern was reported in the literatures [36, 37].

For C15A modified with APTES, the appearance of the similar peaks for planes d_{001} and d_{002} shows that the modification did not alter the MMT characteristics peaks. Previous works have reported an increase in the layer distances, reflected by its d-spacing, by modification with aminosilane [16, 18, 23]. On the contrary, a different phenomenon was observed in this study. This result suggested that the modification mainly occurred at the outside of the clay layers; hence, the d-spacing does not necessarily affect [22]. This is because of the higher hydroxyl content at the broken clay galleries edges has impeded diffusion of the APTES molecules by surfactant from penetrating the clay interlayer. As a result, the chemical reaction with the hydroxyl groups at the surface of the clay is reduced [21]. This assumption carries validity since the diffractogram in Figure 1.6 did not show evidence of d-spacing enlargement.

Interestingly, the d-spacing of the modified C15A was found lower than that of unmodified C15A as demonstrated by the shift of XRD peak towards a higher angle at $2\theta = 3.6°$ (2.45 nm), seemingly different from previous works [21, 22]. The reduction of d-spacing for surface modification of clay seems to originate from several factors. Firstly, the clay plates or silicates layer could not undergo full ion exchange due to the presence of natural defects or charge heterogeneity. As a result, after the ion exchange, some of the organic surfactants may only be physically sorbed onto the surface of the clay instead of ionically bonded. Therefore, during the solvent washes, the surfactant (organic modifier) is easily removed and the plates distance become closer to fill the vacant spacing [8]. Secondly, the counterbalanced reaction by cations from the interlayer space with the negatively charged clay layer causes the surface of the clay to be hydrophilic. This, in turn, increases the difficulty of conventional ion exchange organic molecules intercalation into MMT due to higher hydrophobia properties as well as the bulk of the organic molecules [23]. Similarly, the excess organic modifier may also be removed due to the weaker ionic interaction with the clay.

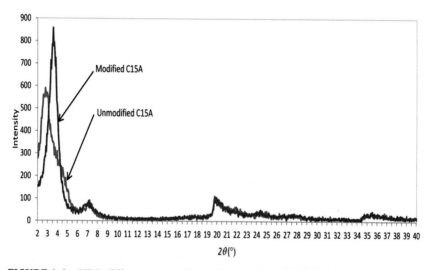

FIGURE 1.6 XRD diffractograms of unmodified and modified C15A.

1.3.3 MORPHOLOGICAL ANALYSIS

The SEM micrographs of both C15As (as received and modified) are shown in Figure 1.7. Generally, the typical size of C15A is 2–13 μm (obtained from manufacturer). However, from this figure, it can be seen that the cluster of clay particles or agglomerates are as large as 40–50 μm. Agglomeration of clay particles is also evident even after modification with APTES. This observation suggests that there is a stronger interaction among the clay particles. These micrographs were further transformed into high contrast images using freeware ImageJ to investigate the effect of silylation on the clay particles morphology. From this image, the estimated average particle size obtained from ImageJ analysis (see Table 1.1) also showed that the modified C15A is smaller than unmodified C15A. The estimated size for unmodified and modified C15A corresponds well with those observed in Figures 1.7a(i) and 7b(i).

Investigations on the effect of C15A modification on polymer-clay interfacial properties were carried by using SEM and FESEM analysis. Cross-sectional SEM micrographs of MMM comprising unsilylated and silylated C15A are shown in Figures 1.8a and 1.8b, respectively. Larger magnification of the interfacial adhesion characteristics were given in the subsequent micrographs in Figure 1.8. From this figure, the differences of the extent of adhesion between the PES and unmodified and modified C15A can be distinguished. For unmodified C15A in Figure 1.8a(ii), it was observed that the clay-polymer interface shows a typical sieve-in-a-cage morphology. This morphology is resulted from poor unmodified C15A-PES matrix compatibility and leads to the interfacial gap formation. Formation of voids was still evident for PES/C15AS0.25, however, the size of the gaps was suppressed than those observed for PES/C15A0.25. This observation suggests that the clay particle was partially functionalized as seen by the fractional 'anchoring' effect on PES/C15AS0.25. However, aminosilane modification provides better polymer coating on the surface of the clay particle.

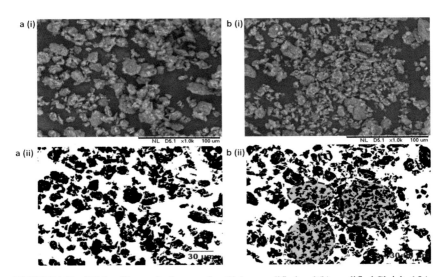

FIGURE 1.7   SEM and ImageJ micrographs of (a) unmodified and (b) modified Cloisite 15A.

TABLE 1.1 ImageJ Analysis of Unmodified and Modified Cloisite15A

Clay	Total Agglomerate Area (µm2)	Average Agglomerate Size (µm)	Area Fraction (%)
Unmodified	17486.286	53.475	44.8
Modified	18201.201	37.762	45.7

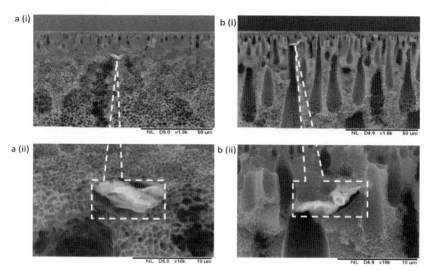

FIGURE 1.8 SEM micrographs for cross section of (a) PES/C15A0.25 and (b) PES/C15AS0.25.

The morphological study was extended with FESEM analysis shown in Figure 1.9 to verify the magnitude of polymer-clay interaction as previously seen in the SEM micrographs. From these micrographs, similar unselective voids were more apparent for PES/C15A0.25 (indicated by arrow). The revealing submicron gaps surrounding the clay particle is resulted from the poor adhesion of between the surface of the clay and the matrix. Meanwhile, according to Figure 1.9b, PES chains were anchored to the clay surface more effectively for aminosilane silylated C15A compared to unsilylated clay. These micrographs were in accordance to those examined by our FTIR, XRD, and SEM analysis, which supports the partially modification behavior and improved interfacial properties. In addition, the size of the clay particle was ~10 µm for both MMMs in Figures 1.8 and 1.9, which is within the range obtained from the manufacturer. The presence of non-agglomerated clay particle further confirms that the C15A were well dispersed. However, even with a good dispersion, issues with the interface voids were unavoidable.

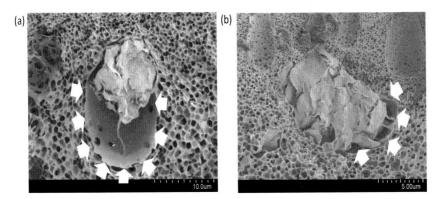

FIGURE 1.9 FESEM micrographs of (a) PES/C15A0.25 and (b) PES/C15AS0.25.

To verify the silylation process of C15A, elemental analysis was performed by EDX for the MMMs and presented in Table 1.2. The elemental compositions corresponding to C15A (Si, Al, and Mg) for PES/C15AS0.25 are two times larger than PES/C15A0.25. This result is suspected to originate from poor distribution of C15A within PES matrix, causing particle localization and provides different elemental composition despite similar filler loadings were incorporated. However, it is worthy to note that the increment for nitrogen was the highest, nearly five times higher than the initial composition. Therefore, it is proposed that the significant increase was due to the nitrogen atoms originated from the attached silane [20], demonstrating the success of the modification of C15A.

TABLE 1.2 Elemental Composition of MMMs Before and After C15A Silylation Analyzed by EDX

Membranes	Atomic Contents (wt%)							
	C	O	S	Si	Al	Mg	N	Total
PES/C15A0.25	66.3	19.1	10.0	3.1	1.3	0.1	0.2	100.0
PES/C15AS0.25	61.3	22.9	6.6	5.5	2.5	0.2	0.9	100.0

Based on Figures 1.8 and 1.9 and Table 1.2, it can be concluded that the aminosilane coupling agent facilitates the polymer-clay adhesion.

From this improvement, we postulated that the higher interfacial area for polymer-clay would contribute to enhanced thermal stability, mechanical properties, and gas separation performance. These changes will be discussed thoroughly in the next sections.

1.3.4   THERMAL ANALYSIS

The decomposition behavior of neat PES and MMMs was examined by thermogravimetric analysis from TGA and DTG (first derivative of TGA) curves shown in Figure 1.10. The thermograms of the neat PES and subsequent MMMs showed two stages of thermal decomposition. TGA analysis reveals that the membranes were thermally stable up to 450°C. The first weight loss of material occurs in the temperature range of 100–220C which corresponds to removal of water molecules physically adsorbed on the surface of the membranes [38]. The second weight loss over a broad range of temperature between 450–700°C attributes to the decomposition of PES polymer chain [39, 40]. In the case of MMMs, the thermograms profile is similar to neat PES below 550°C. Above this temperature, the decomposition profile shows higher thermal stability for PES/C15A0.25 and PES/C15AS0.25 with increased in temperature from 572.5C to 582°C, respectively. This shows that C15A acts as a local heat sink in the polymer matrix, hence the delayed decomposition temperature. In addition, the residual mass of PES/C15A0.25 is 19% higher compared to neat PES due to thermally stable C15A. However, the overall mass loss for PES/C15AS0.25 was observed higher than neat PES and PES/C15A0.25. Compared to all membranes, the TGA curve also shows that the onset decomposition begins the earliest for PES/C15AS0.25, at 145°C. The residual mass is also 12% and 26% lower than neat PES and PES/C15A0.25, respectively. The overall increase in mass loss for PES/C15AS0.25 can be explained by the degradation of intercalated silane and co-occurrence of bonded silane decomposition and dehydroxylation of the clay [16].

The Tg results for neat PES and MMMs are also displayed in Table 1.3. Generally, the change in Tg reflects the polymer chain flexibility of MMMs. A gradual increase in Tg for NCs is expected with the addition of clay particle [41]. However, the membrane's Tg shows a slight reduction after C15A incorporation. This behavior can be associated with the rapid degradation of excess surfactant within the clay layers [42]. It has been reported that most alkyl ammonium surfactants are not stable even at room temperature [11]. For MMMs incorporated with silylated filler, the increase interaction between the surface of the filler and polymer matrix often leads to polymer chain rigidification effect. As a result, these membranes are likely to have higher *Tg* similar to those reported in the literatures [43, 44]. Furthermore, a higher *Tg* is also expected with the removal of some 'unattached' surfactant modifier. As presented in Table 1.3, our results are somewhat different where no obvious increase in the temperature was observed for PES/C15AS0.25. This descent trend is deliberated due to the partial interface interaction between the clay surface and PES matrix which probably outweighs the effect of surfactant removal. SEM and FESEM micrographs in Figures 1.8b and 1.9b have also displayed the partially polymer-clay interface characteristics.

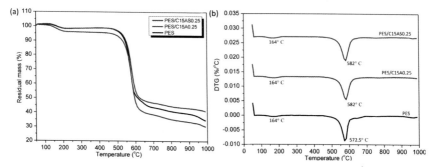

FIGURE 1.10   (a) TGA and (b) DTG curves of neat PES, unmodified, and modified clay membranes.

TABLE 1.3   Glass Transition Temperature, *Tg* Data of PES and Unmodified and Modified 0.25 wt% C15A MMMs

Membrane	Clay Loadings (wt%)	Tg (°C)
PES	0	182.0
PES/C15A0.25	0.25	181.0
PES/C15AS0.25	0.25	181.0

The DSC and TGA results analysis concluded that unmodified C15A incorporation improves the thermal stability of MMM with the delayed degradation temperature while slight reduction of thermal stability was observed for MMMS with modified C15A. Despite the increase interaction near the polymer-clay interface, silylated C15A particle embodiment shows no inhibition of the mobility of polymer chains as the Tg was slightly reduced. In overall, further modification of C15A with aminosilane did not cause severe thermal properties instability of the resultant MMM.

1.3.5   TENSILE TEST

Tensile test was performed in order to investigate the effect of clay addition and aminosilane modification on the overall mechanical properties of the MMM. According to Figure 1.11, the incorporation of C15A enhanced the mechanical strength of the MMMs, as shown by the increase elongation at break and tensile strength relative to neat PES. From this figure, we can conclude that the elongation at break increases in the order of PES < PES/C15A0.25 < PES/C15S0.25. These improvements showed the reinforcing effects of the clay platelets as fillers. A similar observation was reported by Chen et al. [22]. Although the overall improvement was evident, the slight decrease in the tensile strength of PES/C15A0.25 MMM is likely due to the poor interfacial adhesion of the clay particles and polymer matrix as seen in the previous SEM micrographs. Additionally, higher tensile strength was observed for PES/C15AS0.25 after further modification of C15A with APTES. During modification with APTES, the reaction between the silanol groups and amino functional groups on the silicate layers improves the hydrophobicity. This, in turn, enhances the wetting properties at the silicate layers and PES interface [22]. As a result, the induction of chemical reaction between the amino groups from TFC and the tail group from PES further increases the interaction within the components of the MMMs [20]. This behavior confirms that silane is a good adhesion promoter for MMMs [45]. However, since the changes were minimal, we further concluded that the modification reaction was partially successful.

1.3.6   GAS PERMEATION TEST

In order to assess the feasibility of the aminosilane functionalized C15A as filler for MMM (PES/C15AS0.25), gas permeability and selectivity for CO_2 and CH_4 separation were evaluated. These results were compared with neat PES and PES/C15A0.25 gas separation properties and presented in

Figure 1.12. From this figure, for both MMMs, an overall increase in CH_4 permeance was observed in comparison to that neat PES. In addition, CO_2 permeance was also increased by 28.4% for PES/C15A0.25 and decreased marginally around 9.8% for PES/C15AS0.15A. Generally, the increasing permeance trend for PES/C15A0.25 for both gases suggested a typical case of voidage interfacial. These results were also in accordance with the SEM and FESEM micrographs shown in Figures 1.7 and 1.8. Interestingly, silylation of C15A shows prominent changes in gas permeation compared to unmodified C15A when embedded into PES matrix. PES/C15AS0.25 possesses 36.8% and 29.7% in CH_4 and CO_2 permeance lower than PES/ C15A0.25. This could be attributed by the higher interaction between the clay surface and polymer matrix, blocking the easier diffusion path for the gas hence the lower permeance. These claims were in agreement with the partially 'anchored' clay surface in the SEM micrographs in Figure 1.7b(ii) and the improvement in the mechanical properties for PES/C15AS0.25A. Since the Tg data for both MMMs are similar, this further confirms that the interfacial voids effect was more dominant on gas permeation compared to the reduction on polymer chain relaxation.

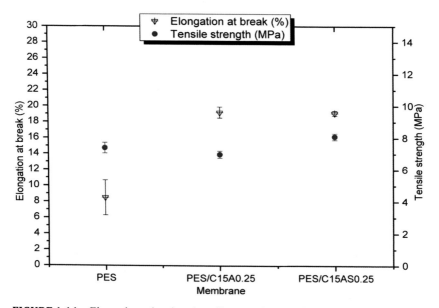

FIGURE 1.11   Elongation at break and tensile strength properties of neat PES and MMMs.

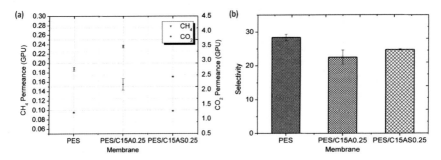

FIGURE 1.12   Gas permeation properties of neat PES and MMMs.

As seen in Figure 1.12b, selectivity of MMM showed a complex trend, 20.7% (PES/C15A0.25) and 12.8% (PES/C15S0.25) lower than neat PES. As previously, discussed, interfacial voids resulted the high permeance for CO$_2$ and CH$_4$ due to negligible transport resistance. However, the lack of the transport resistance on the voids sacrificed the selectivity for the MMMs. Furthermore, the lack of enhancement of the selectivity for PES/C15AS0.25 could also be associated with the proportional and significant reduction of both CO$_2$ and CH$_4$ permeance depicted in Figure 1.12a. It is known that platelet type morphology of C15A inside the PES matrix increased the diffusion path for the gas travel, hence the low permeance is expected [4, 37, 38]. For the present work, it is obvious that the selectivity improvement of PES/C15AS0.25) when compared to PES/C15A0.25 was minimal. As discussed earlier, the formation of interfacial voids was apparent more for PES/C15A0.25, while only minimal interfacial voids was observed from the partial modification of C15A (Figure 1.9). However, this improvement indicates that higher interfacial interaction between the clay surface and polymer matrix may contribute to higher performance of MMM. Nevertheless, fabricating MMM with clay as barriers would still be challenging due to the trade-off relationship between the permeability and selectivity even for the case of successful modification and significant interfacial interaction [46]. In overall, the low selectivity of MMMs in this work compared to neat PES can be correlated with the extent of interfacial defects. Comparing to our recent work [47], we observed that controlling the exfoliation state of the silicate layers gives significant improvement in permeance and selectivity, rather than lowering the interfacial defects as in the present study.

Variation in relative permeance (P_{MMM}/P_{Neat}) and selectivity ($\alpha_{MMM}/\alpha_{Neat}$) for PES/C15A0.25 and PES/C15AS0.25 was further compared in the morphological diagram in Figure 1.13. From this figure, PES/C15A0.25 falls within sieve-in-a-cage morphology (Case 0) as previously described in literature [48]. This is a direct case to diagnose since relative permeance for both gases was obviously higher than that of neat PES and PES/C15A0.25 (see Figure 1.12a). The passing of gas through unselective 'channel' or cages at the interphase results in an overall increase in CO_2 and CH_4 permeance, reducing the selectivity and this is well represented by Figure 1.12b. PES/C15AS0.25, on the other hand, is slightly difficult to interpret. Although it seems that PES/C15AS0.25 has a characteristics of pore-clogging morphology, case IV & V are commonly associated with porous filler, for example, zeolites [48, 49]. As in this study, since the clay is nonporous filler, clogging of pores may not be appropriate to describe the phenomenon. Instead, the reduction of relative permeance can be associated with the tortuosity factor of the high aspect ratio of silicate layers and barrier properties [4, 50]. Although one may presume potential sorption of gas into clay particles may occur [41], previous work by Chung et al. showed that coupling agent (3-aminopropyl)-diethoxymethyl silane (APDEMS) reduced the filler blockage and induced gas transport resistance thus this issue can be ruled out [49].

In order to show the effect of silylation on the MMM, permeance, and selectivity data of PES/C15AS0.25 were plotted in morphological diagram relative to PES. C15A0.25 as shown in Figure 1.13b. This figure shows that the data is scattered at the top left quadrant which corresponds to Case I (matrix rigidification), a typical results of silylation of particles in MMM and improved bonding properties [43]. In addition, the Tg for PES/C15AS0.25 was not severely affected hence the rigidified region for PES/C15AS0.25 was minimal. This is probably due to the partial modification on the clay surface. However, the improvements on the mechanical properties of PES/C15AS0.25 as well as interfacial adhesion as seen in SEM and FESEM micrographs (see Figures 1.7b(ii) and 1.8b) strengthen the claims on the benefits of silylation. From the gas permeation results, it is evident that controlling the severity of interfacial voids may help to tune the selectivity favorably, to a certain extent.

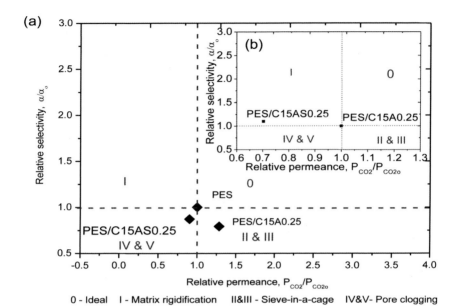

FIGURE 1.13 Morphological diagrams for MMMs.

1.4 CONCLUSIONS

This study prepared twice functionalized organoclay (TFC) by introducing APTES reacting compounds to the hydroxyl groups on the surface of C15A. The addition of functional groups on C15A through silane modification changes the interfacial behavior of polymer-clay and various properties of the MMMs. FTIR-ATR spectras and XRD diffractograms revealed that the surface of the clay was partially modified possibly due to non-rigorous modification environment. SEM and FESEM micrographs of the respective MMMs also showed similar observations. The significant increase in nitrogen content in PES/C15AS0.25 compared to PES/C15A0.25 through EDX elemental analysis supports the presence of the silane moieties. Regardless of the partial modification, incorporation of TFC in MMMs enhanced both the elongation at break and tensile strengths because of the reinforcement effect due to better interfacial interaction. While thermal properties were not severely affected, the improved CO_2/CH_4 separation properties of PES/C15AS0.25 showed that the APTES functional groups were able to provide the 'anchoring' effect by reducing the interface voids distance of the polymer-clay. The enhanced MMM separation performance was obtained by controlling the extent of the polymer-clay interaction and the overall changes on the properties of the membranes. Hence, the use of a silane modifier for surfactant modified clay was able to enhance the compatibility between the polymer and the surface of the clay. For future works, variations of clay modification parameters such as reaction time, initial silane concentration, different functional groups, acid treatment, temperature, solvent types can be investigated to improve the degree of clay silylation [13, 29].

ACKNOWLEDGMENTS

This work was supported by Grant Q.J130000.2542.04H71 and Q.J130000. 3009.00M02 from the University Technology Malaysia through the Research University Grant. The support of the PhD scholarship from the Ministry of Higher Education (MOHE) Malaysia for the author is gratefully acknowledged.

KEYWORDS

- **cloisite15A**
- **gas permeation**
- **mixed matrix membrane**
- **organo alkoxysilanes**
- **polyethersulfone**
- **silane coupling**

REFERENCES

1. Paul, D. R., & Robeson, L. M., (2008). Polymer nanotechnology: Nanocomposites. *Polymer (Guildf), 49*, 3187–3204. doi: 10.1016/j.polymer.2008.04.017.
2. Kotal, M., & Bhowmick, A. K., (2015). Polymer nanocomposites from modified clays: Recent advances and challenges. *Prog. Polym. Sci., 51*, 127–187. doi: 10.1016/j. progpolymsci.2015.10.001.

3. Sciascia, L., Casella, S., Cavallaro, G., Lazzara, G., Milioto, S., Princivalle, F., & Parisi, F., (2019). Olive mill wastewaters decontamination based on organo-nano-clay composites. *Ceram. Int., 45,* 2751–2759. doi: 10.1016/j.ceramint.2018.08.155.

4. Herrera-Alonso, J. M., Marand, E., Little, J. C., & Cox, S. S., (2009). Transport properties in polyurethane/clay nanocomposites as barrier materials: Effect of processing conditions. *J. Memb. Sci., 337,* 208–214. doi: 10.1016/j.memsci.2009.03.045.

5. Picard, E., Vermogen, A., Gerard, J., & Espuche, E., (2007). Barrier properties of nylon 6-montmorillonite nanocomposite membranes prepared by melt blending: Influence of the clay content and dispersion state consequences on modeling. *J. Memb. Sci., 292,* 133–144. doi: 10.1016/j.memsci.2007.01.030.

6. Hashemifard, S. A., Ismail, A. F., & Matsuura, T., (2011). Effects of montmorillonite nano-clay fillers on PEI mixed matrix membrane for CO$_2$ removal. *Chem. Eng. J., 170,* 316–325. doi: 10.1016/j.cej.2011.03.063.

7. Gilman, J. W., Harris, R. H., Shields, J. R., Kashiwagi, T., & Morgan, A. B., (2006). A study of the flammability reduction mechanism of polystyrene-layered silicate nanocomposite: Layered silicate reinforced carbonaceous char. *Polym. Adv. Technol., 17,* 263–271. doi: 10.1002/pat.682.

8. Morgan, A. B., & Harris, J. D., (2003). Effects of organoclay Soxhlet extraction on mechanical properties, flammability properties and organoclay dispersion of polypropylene nanocomposites. *Polymer (Guildf), 44,* 2313–2320. doi: 10.1016/S0032-3861(03)00095-8.

9. Adrar, S., Habi, A., Ajji, A., & Grohens, Y., (2017). Combined effect of epoxy functionalized graphene and organomontmorillonites on the morphology, rheological and thermal properties of poly (butylenes adipate-co-terephtalate) with or without a compatibilizer. *Appl. Clay Sci., 146,* 306–315. doi: 10.1016/j.clay.2017.06.009.

10. Adrar, S., Habi, A., Ajji, A., & Grohens, Y., (2018). Synergistic effects in epoxy functionalized graphene and modified organo-montmorillonite PLA/PBAT blends. *Appl. Clay Sci., 157,* 65–75. doi: 10.1016/j.clay.2018.02.028.

11. Han, H., Rafiq, M. K., Zhou, T., Xu, R., Mašek, O., & Li, X., (2019). A critical review of clay-based composites with enhanced adsorption performance for metal and organic pollutants. *J. Hazard. Mater., 369,* 780–796. doi: 10.1016/j.jhazmat.2019.02.003.

12. Sinha, R. S., & Okamoto, M., (2003). Polymer/layered silicate nanocomposites: A review from preparation to processing. *Prog. Polym. Sci., 28,* 1539–1641. doi: 10.1016/j.progpolymsci.2003.08.002.

13. He, H., Tao, Q., Zhu, J., Yuan, P., Shen, W., & Yang, S., (2013). Silylation of clay mineral surfaces. *Appl. Clay Sci., 71,* 15–20. doi: 10.1016/j.clay.2012.09.028.

14. Vazquez, A., López, M., Kortaberria, G., Martín, L., & Mondragon, I., (2008). Modification of montmorillonite with cationic surfactants. Thermal and chemical analysis including CEC determination. *Appl. Clay Sci., 41,* 24–36. doi: 10.1016/j.clay.2007.10.001.

15. Pavlidou, S., & Papaspyrides, C. D., (2008). A review on polymer-layered silicate nanocomposites. *Prog. Polym. Sci., 33,* 1119–1198. doi: 10.1016/j.progpolymsci.2008.07.008.

16. Huskić, M., Zigon, M., & Ivanković, M., (2013). Comparison of the properties of clay polymer nanocomposites prepared by montmorillonite modified by silane and by quaternary ammonium salts. *Appl. Clay Sci., 85,* 109–115. doi: 10.1016/j.clay. 2013.09.004.

17. Bruce, A. N., Lieber, D., Hua, I., & Howarter, J. A., (2014). Rational interface design of epoxy-organoclay nanocomposites: Role of structure-property relationship for silane modifiers. *J. Colloid Interface Sci., 419*, 73–78. doi: 10.1016/j.jcis.2013.12.051.

18. Piscitelli, F., Posocco, P., Toth, R., Fermeglia, M., Pricl, S., Mensitieri, G., & Lavorgna, M., (2010). Sodium montmorillonite silylation: Unexpected effect of the aminosilane chain length. *J. Colloid Interface Sci., 351*, 108–115. doi: 10.1016/j.jcis.2010.07.059.

19. Shanmugharaj, A. M., Rhee, K. Y., & Ryu, S. H., (2006). Influence of dispersing medium on grafting of aminopropyltriethoxysilane in swelling clay materials. *J. Colloid Interface Sci., 298*, 854–859. doi: 10.1016/j.jcis.2005.12.049.

20. Kim, E. S., Shim, J. H., Woo, J. Y., Yoo, K. S., & Yoon, J. S., (2010). Effect of the silane modification of clay on the tensile properties of nylon 6/clay nanocomposites. *J. Appl. Polym. Sci., 117*, 809–816. doi: 10.1002/app.

21. Joo, J. H., Shim, J. H., Choi, J. H., Choi, C. H., Kim, D. S., & Yoon, J. S., (2008). Effect of silane modification of an organoclay on the properties of polypropylene/clay composites. *J. Appl. Polym. Sci., 109*, 3645–3650. doi: 10.1002/app.

22. Chen, G. X., Kim, H. S., Shim, J. H., & Yoon, J. S., (2005). Role of epoxy groups on clay surface in the improvement of morphology of poly(L-lactide)/clay composites. *Macromolecules, 38*, 3738–3744. doi: 10.1021/ma0488515.

23. Shen, W., He, H., Zhu, J., Yuan, P., & Frost, R. L., (2007). Grafting of montmorillonite with different functional silanes via two different reaction systems. *J. Colloid Interface Sci., 313*, 268–73. doi: 10.1016/j.jcis.2007.04.029.

24. Kusworo, T. D., Ismail, A. F., Mustafa, A., & Matsuura, T., (2008). Dependence of membrane morphology and performance on preparation conditions: The shear rate effect in membrane casting. *Sep. Purif. Technol., 61*, 249–257. doi: 10.1016/j.seppur.2007.10.017.

25. Cervantes-Uc, J. M., Cauich-Rodríguez, J. V., Vázquez-Torres, H., Garfias-Mesías, L. F., & Paul, D. R., (2007). Thermal degradation of commercially available organoclays studied by TGA-FTIR. *Thermochim. Acta, 457*, 92–102. doi: 10.1016/j.tca.2007.03.008.

26. Mokhtar, N. M., Lau, W. J., Ismail, A. F., & Ng, B. C., (2014). Physicochemical study of polyvinylidene fluoride-Cloisite15A® composite membranes for membrane distillation application. *RSC Adv., 4*, 63367–63379. doi: 10.1039/C4RA10289D.

27. Bhole, Y. S., Wanjale, S. D., Kharul, U. K., & Jog, J. P., (2007). Assessing feasibility of polyarylate-clay nanocomposites towards improvement of gas selectivity. *J. Membr. Sci., 306*, 277–286. doi: 10.1016/j.memsci.2007.09.001.

28. Chen, G. X., Choi, J. B., & Yoon, J. S., (2005). The role of functional group on the exfoliation of clay in poly(L-lactide). *Macromol. Rapid Commun., 26*, 183–187. doi: 10.1002/marc.200400452.

29. Herrera, N. N., Letoffe, J. M., Reymond, J. P., & Bourgeat-Lami, E., (2005). Silylation of laponite clay particles with monofunctional and trifunctional vinyl alkoxysilanes. *J. Mater. Chem., 15*, 863–871. doi: 10.1039/b415618h.

30. Yang, S., Yuan, P., He, H., Qin, Z., Zhou, Q., Zhu, J., & Liu, D., (2012). Effect of reaction temperature on grafting of γ-aminopropyl triethoxysilane (APTES) onto kaolinite. *Appl. Clay Sci., 62–63*, 8–14. doi: 10.1016/j.clay.2012.04.006.

31. Han, J., Lee, W., Choi, J. M., Patel, R., & Min, B. R., (2010). Characterization of polyethersulfone/polyimide blend membranes prepared by a dry/wet phase inversion:

Precipitation kinetics, morphology and gas separation. *J. Memb. Sci., 351*, 141–148. doi: 10.1016/j.memsci.2010.01.038.

32. Ge, L., Zhu, Z., & Rudolph, V., (2011). Enhanced gas permeability by fabricating functionalized multi-walled carbon nanotubes and polyethersulfone nanocomposite membrane. *Sep. Purif. Technol., 78*, 76–82. doi: 10.1016/j.seppur.2011.01.024.

33. Sorrentino, A., Tortora, M., & Vittoria, V., (2006). Diffusion behavior in polymer-clay nanocomposites. *J. Polym. Sci. Part B Polym. Phys., 44*, 265–274. doi: 10.1002/polb.20684.

34. Azeez, A. A., Rhee, K. Y., Park, S. J., & Hui, D., (2013). Epoxy clay nanocomposites-processing, properties and applications: A review. *Compos. Part B Eng., 45*, 308–320. doi: 10.1016/j.compositesb.2012.04.012.

35. Villaluenga, J. P. G., Khayet, M., López-Manchado, M. A., Valentin, J. L., Seoane, B., & Mengual, J. I., (2007). Gas transport properties of polypropylene/clay composite membranes. *Eur. Polym. J., 43*, 1132–1143. doi: 10.1016/j.eurpolymj.2007.01.018.

36. Jaafar, J., Ismail, A. F., & Matsuura, T., (2009). Preparation and barrier properties of SPEEK/Cloisite 15A®/TAP nanocomposite membrane for DMFC application. *J. Membr. Sci., 345*, 119–127. doi: 10.1016/j.memsci.2009.08.035.

37. Zulhairun, A. K., Ismail, A. F., Matsuura, T., Abdullah, M. S., & Mustafa, A., (2014). Asymmetric mixed matrix membrane incorporating organically modified clay particle for gas separation. *Chem. Eng. J., 241*, 495–503. doi: 10.1016/j.cej.2013.10.042.

38. Herrera-Alonso, J. M., Sedlakova, Z., & Marand, E., (2010). Gas barrier properties of nanocomposites based on in situ polymerized poly(n-butyl methacrylate) in the presence of surface modified montmorillonite. *J. Memb. Sci., 349*, 251–257. doi: 10.1016/j.memsci.2009.11.057.

39. Adib, H., Hassanajili, S., Mowla, D., & Esmaeilzadeh, F., (2015). Fabrication of integrally skinned asymmetric membranes based on nanocomposite polyethersulfone by supercritical CO_2 for gas separation. *J. Supercrit. Fluids., 97*, 6–15. doi: 10.1016/j.supflu.2014.11.001.

40. Defontaine, G., Barichard, A., Letaief, S., Feng, C., Matsuura, T., & Detellier, C., (2010). Nanoporous polymer-clay hybrid membranes for gas separation. *J. Colloid Interface Sci., 343*, 622–627. doi: 10.1016/j.jcis.2009.11.048.

41. Liang, C. Y., Uchytil, P., Petrychkovych, R., Lai, Y. C., Friess, K., Sipek, M., Mohan, R. M., & Suen, S. Y., (2012). A comparison on gas separation between PES (polyethersulfone)/MMT (Na-montmorillonite) and PES/TiO_2 mixed matrix membranes. *Sep. Purif. Technol., 92*, 57–63. doi: 10.1016/j.seppur.2012.03.016.

42. Cui, L., Khramov, D. M., Bielawski, C. W., Hunter, D. L., Yoon, P. J., & Paul, D. R., (2008). Effect of organoclay purity and degradation on nanocomposite performance, Part 1: Surfactant degradation. *Polymer (Guildf), 49*, 3751–3761. doi: 10.1016/j.polymer.2008.06.029.

43. Ismail, A. F., Kusworo, T., & Mustafa, A., (2008). Enhanced gas permeation performance of polyethersulfone mixed matrix hollow fiber membranes using novel dynasylan Ameo silane agent. *J. Memb. Sci., 319*, 306–312. doi: 10.1016/j.memsci.2008.03.067.

44. Li, F., Li, Y., Chung, T. S., & Kawi, S., (2010). Facilitated transport by hybrid POSS®–Matrimid®–Zn2+ nanocomposite membranes for the separation of natural gas. *J. Memb. Sci., 356*, 14–21. doi: 10.1016/j.memsci.2010.03.021.

45. Wang, Z., Suo, J., & Li, J., (2009). Synthesis and characterization of epoxy resin modified with γ-thiopropyl triethoxy silane. *J. Appl. Polym. Sci., 114*, 2388–2394. doi: 10.1002/app.

46. Robeson, L. M., (2008). The upper bound revisited. *J. Memb. Sci., 320,* 390–400. doi: 10.1016/j.memsci.2008.04.030.

47. Ismail, N. M., Ismail, A. F., Mustafa, A., Matsuura, T., Soga, T., Nagata, K., & Asaka, T., (2015). Qualitative and quantitative analysis of intercalated and exfoliated silicate layers in asymmetric polyethersulfone/cloisite15A® mixed matrix membrane for CO2/CH4 separation. *Chem. Eng. J., 268*, 371–383. doi: 10.1016/j.cej.2014.11.147.

48. Moore, T. T., & Koros, W. J., (2005). Non-ideal effects in organic-inorganic materials for gas separation membranes. *J. Mol. Struct., 739*, 87–98. doi: 10.1016/j.molstruc.2004.05.043.

49. Li, Y., Guan, H. M., Chung, T. S., & Kulprathipanja, S., (2006). Effects of novel silane modification of zeolite surface on polymer chain rigidification and partial pore blockage in polyethersulfone (PES)-zeolite A mixed matrix membranes. *J. Memb. Sci., 275*, 17–28. doi: 10.1016/j.memsci.2005.08.015.

50. Goodarzi, V., Hassan, J. S., Ali Khonakdar, H., Ghalei, B., & Mortazavi, M., (2013). Assessment of role of morphology in gas permselectivity of membranes based on polypropylene/ethylene vinyl acetate/clay nanocomposite. *J. Membr. Sci., 445*, 76–87. doi: 10.1016/j.memsci.2013.04.073.

CHAPTER 2

Photocatalytic Activity of ZnO-PEG Nanoparticles for Palm Oil Mill Secondary Effluent (POMSE) Treatment

NUR HANIS HAYATI HAIROM,[1] NURUL HANA ISMAIL,[1]
NURUL AISYAH ABD HADI,[1] DILAELYANA ABU BAKAR SIDIK,[1,2] and
ABDUL WAHAB MOHAMMAD[3]

[1]*Faculty of Engineering Technology, University Tun Hussein Onn Malaysia, 86400 Parit Raja, Batu Pahat, Johor, Malaysia, E-mail: nhanis@uthm.edu.my (N. H. H. Hairom)*

[2]*Center of Diploma Studies, University Tun Hussein Onn Malaysia, 86400 Parit Raja, Batu Pahat, Johor, Malaysia*

[3]*Center for Sustainable Process Technology (CESPRO), Faculty of Engineering and Built Environment, University Kebangsaan Malaysia, 43600 UKM Bangi, Selangor, Malaysia*

ABSTRACT

Palm oil mill secondary effluent (POMSE) has a high color intensity, dissolve oxygen, turbidity, and an organic load of BOD which still not achieved the discharged requirement by the Department of Environment (DOE) and led to detrimental to the aquatic life. The photocatalytic degradation process is one of the promising methods in wastewater treatment due to its advantages. However, the study on POMSE treatment using the photocatalytic degradation process in the presence of ZnO-PEG nanoparticles (NPs) is still limited. Therefore, this study reports on the photocatalytic degradation of POMSE by using ZnO-PEG NPs. The ZnO-PEG NPs was characterized by using XRD and FTIR where the results show that there are no impurities present in the samples and presenting the nature and chemical bonds of ZnO-PEG nanoparticle. Then, the optimization of the photocatalytic degradation of POMSE in a UV-activated ZnO system based on central composite design (CCD) in response surface methodology (RSM) was determined. ZnO-PEG NPs have a great potential in degradation of POMSE and this is supported with the results obtained from the experimental works. Three potential factors which are initial pH of POMSE (A), Loading of ZnO-PEG (B), and concentration of POMSE (C) were evaluated for the significance design of experiment. It is found that all the three main factors were significant, with contributions of 34.5% (A), 79% (B), and 82% (C) respectively, to the POMSE degradation. Accordingly, the optimum condition for the photocatalysis degradation process of POMSE is under pH 6.5 in presence of ZnO-PEG with 0.08 g/L for the 25% of POMSE concentration. Then, the photocatalytic activity mechanism of ZnO-PEG nanoparticle was studied by using the kinetic study. It is believed that this integrated approach can be implemented in the industry to achieve discharged standard of POMSE and maintain the green environment for future generation.

2.1   INTRODUCTION

Palm oil industry is one of the leading industries in Malaysia with production of 18.79 million tonnes of crude palm oil (CPO) from an oil palm planted area of 5.08 million hectares in 2012 [1]. However, such production has resulted a large amount of palm oil mill effluent (POME), estimated at nearly three times the quantity of CPO. For each tonne of fresh fruit bunches (FFB) processed, an approximate of 0.67 tonne of POME is generated. Hence, in 2012, about 64.29 million tonnes of POME was produced [1]. POME has high organic load (chemical oxygen demand (COD): 45,500–65,000: biological oxygen demand (BOD): 21,500–28,500) which is detrimental to aquatic life [2]. In Malaysia, more than 85% of the millers employ open ponding system for the POME treatment, and enforcement is carried out by the Department of Environment (DOE) that was enacted in the year 1984. Malaysia Environmental Quality states that the safe level for POME waste to be discharged is fixed at 100 ppm of BOD level, while the COD level must be reduced down to 50 ppm before being discharged into the waterway [3]. However, the existing ponding treatment failed to meet the permitted level for discharge where the BOD and COD level are still higher than the discharge threshold [4]. Therefore, the POME that has been treated by using the method like the ponding treatment will be continued to be treated by the other proposed method. This POME is now known as palm oil mill secondary effluent (POMSE) as they have been treated at the first stage. Although the POMSE has been through treatment, the characteristics of POMSE still not achieved the discharged requirement by the DOE where the values of the COD, BOD as well as the color intensity still high and considered as dangerous to be discharged [5]. The reduction of oils and grease, however, is seen to be significantly reduced at the primary treatment. Therefore, another treatment is proposed in this study to solve this problem which is the photocatalytic degradation process.

Recent study Ng and Cheng [4] show that POME photo-treatment using ultraviolet (UV) activated photocatalysts leads to another interesting pathway. In 20 hours of UV exposure, around 80% of the organics in the POME was photomineralized significantly. This process is known as photocatalytic degradation and considered as an emerging technology of dye wastewater treatment. This process is promising as in the past decades; photocatalytic techniques have been shown many advantages over the traditional technique including quick oxidation, no formation of polycyclic products and oxidation of pollutants up to the parts per billion (ppb) levels [6]. However, the POMSE treatment using photocatalytic degradation process is still limited. Different nanoparticles (NPs) have been used in photocatalysis studies such as titanium dioxide (TiO_2), zinc oxide (ZnO), tin dioxide (SnO_2), cadmium sulfide (CdS), iron oxide (Fe_2O_3), etc. for various wastewater treatments [7]. Among these NPs, ZnO has been claimed as the most effective photocatalyst. When the photocatalytic degradation by ZnO is applied, hydroxyl radicals ($^\bullet$OH) are generated as soon as the photocatalyst is illuminated by the UV light [8]. These radicals then attacked the organic compounds in the bulk solution and consequently results in the degradation of the POMSE.

Recently, Chiranont [9] confirmed that the industrial newsprint wastewater was successfully treated by photocatalytic degradation process with utilization of self-synthesized zinc oxide nanoparticles (ZnO-PEG). However, the characterization of ZnO-PEG and its application for POMSE treatment has not been investigated. Therefore, this proposed research aims to characterize the ZnO-PEG NPs in order to identify its chemical bonds, crystallinity, and purity. Consequently, ZnO-PEG NPs will be used for the photocatalytic degradation of POMSE under various parameters. It is believed that the standard discharge value can be achieved via photocatalytic degradation process for the industrial POMSE treatment.

The treatment of the industrial POMSE via photocatalytic degradation process has not been optimized yet which brings to the fact that the required condition that are suitable to run the experiments on the POMSE is unknown. Therefore, to ensure the POMSE treatment done under the required condition, the response surface methodology (RSM) has been used. According to the previous research by Cheng et al. [5], the application of the central composite design (CCD) under RSM methodology definitely leading to the better prediction of the output response with a significantly reduced experiment runs. The method employed by the CCD is a statistical method used for the factor's optimization. Concurrently, it also evaluates the significance of each factor and their effects on the response [10]. This advantage allows the CCD to determine the important factors as well as the range of the factors needed in the experimental research. In this research which is focusing in treating the POMSE, the important factors are the initial pH of POMSE, concentration of POMSE and ZnO loading. Hence, the RSM is used to determine the desired condition of the factors involved in the POMSE treatment.

2.2   LITERATURE REVIEW

2.2.1   *MECHANISM OF PHOTOCATALYTIC DEGRADATION PROCESS*

One type of ZnO-PEG NPs has been chosen to be used as a photocatalyst (Figure 2.1) in the photocatalytic degradation of wastewater which is 0.015 g/L ZnO-PEG. This is due to this loading was found to be the optimum loading of PEG encapsulated ZnO based on the previous research by Chiranont [9]. Photocatalytic degradation [11] process was started by photoexcitation the ZnO and then followed with the creation of an electron-hole pair on the NPs surface in Eqn. (1):

$$ZnO + hv \rightarrow ZnO\ (e_{cb}^- + h_{vb}^+) \tag{1}$$

The h_{vb}^+ are excellent oxidants that leads to the direct oxidation of the POMSE. The direct oxidation process is shown in the Eqn. (2):

$$h_{vb}^+ + POMSE_{ads} \rightarrow POMSE_{ads}^+ \tag{2}$$

Generation of strong oxidizing •OH radicals were occurred in the indirect oxidation process where the hydroxyl radicals were extremely unstable which then leads to the degradation of the organic particles in the POMSE

[7]. The reaction of the h_{vb}^+ reacted with electron donors like water and hydroxyl ions (OH$^-$) are shown in the Eqns. (3) and (4):

$$h_{vb}^+ + H_2O \rightarrow \bullet OH + H^+ \tag{3}$$

$$h_{vb}^+ + OH^- \rightarrow \bullet OH \tag{4}$$

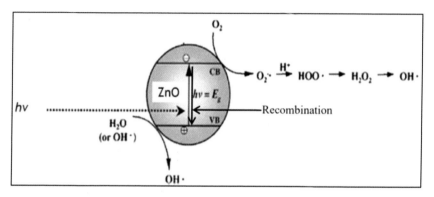

FIGURE 2.1 Schematic diagram displaying the production of oxidative agents via photocatalytic reaction [11].

The oxygen molecules (O_2) are the excellent electron receptors which reduced to the superoxide anion radicals $O_2\bullet$ by the action of the electron in the conduction band, e_{cb}^- as shown in the Eqn. (5):

$$e_{cb}^- + O_2 \rightarrow O_2\bullet^- \tag{5}$$

Organic peroxides may be formed in the by the radicals in the presence of the organic scavenger (Eqn. (6)) or hydrogen peroxide (H_2O_2) (Eqn. (7)):

$$O_2\bullet^- + POMSE \rightarrow POMSE - OO\bullet \tag{6}$$

$$O_2\bullet^- + HO_2^\bullet + H^+ \rightarrow H_2O_2 + O_2 \tag{7}$$

Excess H_2O_2 reacts with hydroxyl radicals and holes to produce $HO_2\bullet$ as shown in Eqns. (8) and (9):

$$H_2O_2 + OH \rightarrow HO_2^\bullet + H_2O \tag{8}$$

$$H_2O_2 + hVB^+ \rightarrow H^+ + HO_2^\bullet \tag{9}$$

Conduction band's electron also contribute in the production of hydroxyl radicals and holes (Eqns. (10) and (11)), which have been indicated as the primary cause of organic matter mineralization (Eqn. (12)).

$$H_2O_2 + eCB^- \rightarrow {}^{\bullet}OH + OH^- \tag{10}$$

$$H_2O_2 + {}^{\bullet}O_2^- \rightarrow {}^{\bullet}OH + OH^- + O_2 \tag{11}$$

$${}^{\bullet}OH + POMSE \rightarrow \text{Degraded POMSE} \tag{12}$$

2.3 METHODOLOGY

2.3.1 WASTEWATER SAMPLING

The POMSE sample was collected at the effluent point of an open pond in a local palm oil located in the state of Johor, Malaysia.

2.3.2 CHARACTERIZATION OF ZNO-PEG NANOPARTICLE

Characterization of the ZnO-PEG NPs were conducted by x-ray diffractometer (XRD) (Bruker AXS GmBH model) and Fourier transform infrared spectroscopy (FTIR) (Perkin-Elmer FTIR (Spectrum RX-1) spectrophotometer (Nicolet FT-IR Avatar 360).

2.3.3 EXPERIMENTAL SET-UP AND OPERATION

Before starting the experiment, Minitab 17 software was used where the RSM specifically design the experiments. There are three potential factors, initial pH of POMSE, loading of ZnO-PEG (g/L) and concentration of POMSE. These three factors were optimized in the current study to maximize the degradation (response %). The experiment was then carried out as in Figure 2.2. The basic part of the apparatus is the conical flask which acts as a photocatalytic reactor in batch method. The UV lamp (253.7 nm, 18 W, GPH 295T5L 4PSE, USA) was placed to activate the photocatalysts. The POMSE sample was poured into the reactor with ZnO-PEG NPs before the photocatalysis process started. 1L of POMSE and ZnO-PEG NPs was well agitated by magnetic stirrer (150 rpm) under UV radiation. The photocatalytic reactor was placed in water bath throughout the experiment to maintain the temperature under 25°C. Hydrochloric acid (HCl) and sodium hydroxide (NaOH) solutions were used to adjust the solution pH that will be measured by pH meter (Eutech Model). At the interval of 5 minutes, 5 ml of degraded wastewater was sampled and the mixed solution (POMSE and ZnO-PEG) will be separated by using centrifuge (Eppendorf, 5804 model) under 500 rpm for 20 minutes.

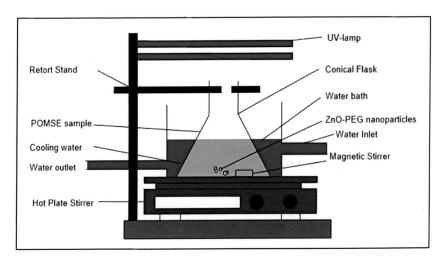

FIGURE 2.2 Experimental set-up of photocatalytic degradation process.

2.3.4 ANALYTICAL METHOD

The treated POMSE was analyzed in term of pH, color intensity, turbidity, dissolved oxygen (DO) and BOD. pH value of the treated POMSE was measured by using the pH meter (Eutech model). Analysis for color intensity was done by using UV-VIS Scanning Spectrophotometer (Spectro UV-2650, Labomed. Inc.). The turbidity of the samples was measured with the turbidity meter (Eutech model, TN-100). While the analysis for the DO and BOD was measured using the DO meter (Hanna Instruments, HI9146).

2.4 RESULTS AND DISCUSSIONS

2.4.1 CHARACTERIZATION OF ZnO-PEG NANOPARTICLES (NPS)

Characterization of ZnO-PEG NPs was conducted by using the X-ray Powder Diffraction (XRD - Bruker AXS GmbH model) to ascertain the purity and crystallinity of the ZnO-PEG NPs that had been synthesized by using the precipitation method. Figure 2.1 shows the XRD pattern of the ZnO-PEG NPs. The diffractogram shows that the ZnO-PEG NPs exhibits sharp diffraction peak characteristics which implies that pure ZnO was formed through the precipitation method. There are no characteristic diffraction peaks from other phases or impurities were detected as can be seen in the pattern indicated that the sample is pure without any other impurities.

The characteristics of ZnO-PEG NPs in terms of nature and chemical bonds were characterized by using the FTIR through the production of an infrared (IR) absorption spectrum. The nature of chemical bonds of ZnO-PEG NPs was shown in Figure 2.4. The absorption peaks were obtained at 3414.14, 1509, 877.555 and 632cm^{-1}. The peaks at 3414.14 cm^{-1} and 877.55 cm^{-1} indicates the O-H stretching and deformation due to the adsorption of moisture from surrounding during synthesizing the photocatalyst. The peak at 1509 cm^{-1} indicates the vibration and stretching of C=O bond while the peak at 632 cm^{-1} are correspond to the Zn-O stretching and deformation vibration (Figures 2.3 and 2.4).

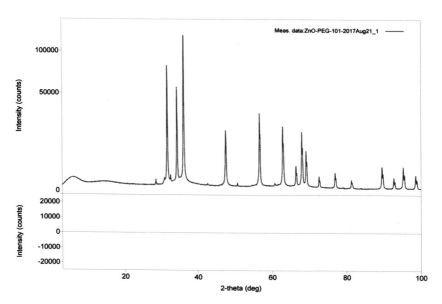

FIGURE 2.3   XRD pattern of ZnO-PEG nanoparticles.

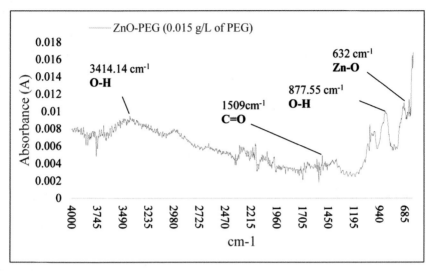

FIGURE 2.4 FTIR image of ZnO-PEG nanoparticle.

2.4.2 EFFECT OF INITIAL PH

In the industrial wastewater treatment, regulation of pH value is important as it is a measure of how acidic or basic water is. The regulation of pH before treating the wastewater is crucial because the treatment process can be harmed by the dangerous level of acidic and basic condition. It affects the charge of the photocatalysts particles, the point at which the conductance and valence bands are located [12]. The value of pH is an important indicator of water that is changing chemically as stated by Idris et al. [13] and crucial characteristics of wastewater because it determines degradation efficiency. The photocatalytic activity of the pH at different pH condition (pH 4, 6.5, and 9) according to the RSM range has been studied to observe the effect of the different pH in the photocatalytic activity. The other parameters which are concentration of POMSE and catalyst loading were fixed at 100% concentration and 0.08 g/L, respectively. Figure 2.5(a) shows the pattern of the POMSE degradation percentage on the color removal under different pH in 60 minutes of duration.

Based on the results obtained from Figure 2.5(a), the photocatalytic activity under pH 6.5 for 100% concentration of POMSE shows the highest percentage of degradation which is 34.5% in contrast to pH 4 (19.1%) and pH 9 (19.2%) in terms of color removal. Figure 2.5(b) presents the percentage of degradation in turbidity during the photocatalytic activity in presence of ZnO-PEG (0.015 g/L of PEG) under different pH. The turbidity of the 100% concentration of POMSE was found to be 543 NTU before treatment and 283 NTU after the treatment respectively. The photocatalytic activity under pH 6.5 shows the highest percentage of turbidity removal (48%) and becomes constant after the 2 minutes of the process. Meanwhile, pH 9 shows the lowest percentage of turbidity removal (19.8%).

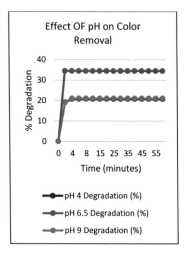 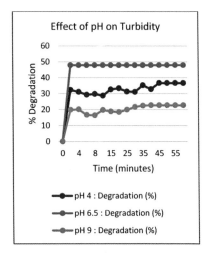

FIGURE 2.5   Percentage of degradation of a) color removal versus time b) turbidity versus time for photocatalysis process in presence of ZnO-PEG under different pH.

Figure 2.6 and Table 2.1 shows the concentration of DO under the influence of different pH. DO refers to the level of free, non-compound oxygen present in water or other liquids and one of the important parameters in assessing water quality because of its influence on the organisms living within the water. A DO level that is too high or too low can harm the aquatic life and affect water quality [12]. Based on Figure 2.5 and Table 2.3, the reading of DO for the samples under different pH is observed to be at 5.78 ppm at pH 4 and 4.93 ppm at pH 9, while the DO value for the optimum pH is at 6.09 ppm. DO value under the optimum pH 6.5 had the highest reading which is 6.09 ppm which shows the most desired reading. DO value under 5.0 ppm consider as harmful to the aquatic life so, 6.09 ppm is the safe value for the water to be discharged to the river. From the analyzed results, it can be concluded that the treated POMSE is safe to be discharged to the environment since the DO value is within the acceptable range.

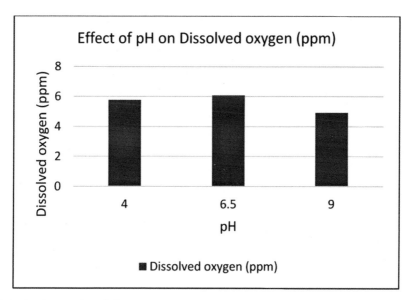

FIGURE 2.6   Value of dissolved oxygen (DO) for photocatalysis process in presence of ZnO-PEG (0.015 g/L) under different pH for highest degradation time.

TABLE 2.1   Characteristics of 100% Concentration of POMSE After Photocatalytic Degradation in the Presence of 0.08 g/L ZnO-PEG Under Different pH

Solution pH	POMSE Degradation (%)	Turbidity Reduction (%)	Dissolved Oxygen (ppm)	BOD$_5$
4	20.8	33.6	5.78	93
6.5	34.5	47.9	6.09	72
9	21.1	22.7	4.93	93

The results clearly show that the pH significantly affects the photocatalytic activity efficiency. According to the previous study of POMSE by Alhaj et al. [12], POMSE have been degraded efficiently at the pH near to the neutral condition. It has been established that at lower pH, the functional groups are protonated, thus raising the positive charge of the photocatalyst surface which decreases the degradation of organic molecules. While high value of pH of a solution shows the reduction on the degradation of organic molecules as the hydroxyl ions compete with the organic molecules for the adsorption on the surface of the catalysts. Thus, degradation efficiency declined at the lower pH or acidic condition and it can be deduced that the optimum pH for the photocatalytic activity of ZnO-PEG (0.015 g/L) is at pH 6.5.

2.4.3 EFFECT OF CATALYST LOADING

Effect of ZnO-PEG loading from 0.08 to 0.90 g/L for photocatalytic degradation of POMSE was investigated under pH 6.5 with a fixed concentration of POMSE (25%). It can be observed that in Figure 2.6(a) the color intensity of the treated POMSE decreased significantly when the loading of ZnO-PEG at its lowest. The amount of ZnO-PEG needed to affect a photocatalytic reaction grossly affects the overall process of photo-degradation just as its concentration is equally important to ensure a true heterogeneous photocatalytic system [14]. The high amount of loading increasing the photocatalytic activity, however, once a saturation phase had been obtained; the increase in the loading of the catalyst causes cloudiness in the POMSE. There has been a radial corresponding decrease in the efficiency of light photon adsorption due to the cloudiness caused by the additional catalyst and this further leads to the decrease in a surface area exposed to irradiation and thus decreases the photocatalytic effectiveness of the process [15].

The effect of the different catalyst loading can be observed where 0.08 g/L shows the highest degradation percentage (79%) in contrast to 0.50 g/L and 0.90 g/L (78% and 77%) though there are only small differences. Based on Figure 2.7(a) and Table 2.2, it can be analyzed that the most effective loading of a catalyst in the photocatalytic activity is 0.08 g/L. Figure 2.7(b) shows the trend of the turbidity removal percentage under different loading of photocatalysts (0.08 g/L, 0.50 g/L and 0.90 g/L). The percentage of turbidity removal of POMSE for different loading of ZnO-PEG (0.08 g/L, 0.50 g/L and 0.90 g/L) are 82.5%, 68.7%, and 73.1%. The highest turbidity removal obtained was for the loading of 0.08 g/L and this is corresponding to the percentage of color removal. Therefore, the greatest turbidity removal of POMSE obtained was for 0.08 g/L of ZnO-PEG.

Figure 2.8 demonstrates the pattern of DO value obtained for different loading of ZnO-PEG (0.08 g/L, 0.50 g/L and 0.90 g/L). based on Figure 2.8, the DO value of different loading of photocatalysts (0.08 g/L, 0.50 g/L and 0.90 g/L) obtained for the highest degradation time are 6.64 ppm, 6.98 ppm, and 6.35 ppm. All the obtained values are under the desired value which is higher than 5 ppm in order to provide a better living environment to the aquatic life. This support the finding on the previous research by Chong et al. [14], where increasing the amount of catalyst reduce the degradation efficiency as there are decrease in the electron-hole produced by the light photon. Hence, the optimum amount of ZnO-PEG loading is found to be 0.08 g/L.

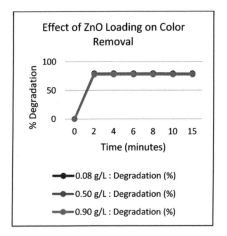 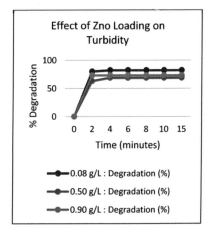

FIGURE 2.7 Percentage of degradation on a) color removal versus time b) Turbidity versus time for photocatalysis process in presence of ZnO-PEG (0.015 g/L) under different loading of ZnO-PEG.

TABLE 2.2 Characteristics of 25% Concentration of POMSE After Photocatalytic Degradation in the Presence of Different Loading of ZnO-PEG

Loading of ZnO-PEG (g/L)	POMSE Degradation (%)	Turbidity Reduction (%)	Dissolved Oxygen (ppm)	BOD$_5$
0.08	79.0	82.5	6.64	59
0.50	78.3	68.7	6.98	74
0.90	77.5	73.1	6.35	94

2.4.4 EFFECT OF POMSE CONCENTRATION

The concentration of POMSE is one of the important factors which affecting the effectiveness of the photocatalytic degradation process. This is due to the nature and concentration of organic constituents present in the POMSE where the higher the concentration of POMSE, the higher the concentration of organic constituents in the POMSE. If the surface of the photocatalyst is highly saturated with the concentration of the pollutants, there will be a reduction on the photonic efficiency and the catalyst becomes deactivated [16]. For the 25% and 50% of the POMSE concentration, the water was added to dilute the POMSE. This experiment was done under optimum pH (pH 6.5) and optimum loading (0.08 g/L) based on previous findings. Figure 2.9(a) shows the percentage of dye degradation under different concentration of POMSE. The results obtained for the three different concentrations (25%, 50%, and 100%) are 79%, 55%, and 34.5%, respectively. This shows that at the lowest concentration of POMSE, the highest percentage of POMSE degradation is obtained. While at 100% concentration of POMSE, the percentage of degradation shows the lowest result. Turbidity level of the POMSE was also been investigated to observe the effect of the POMSE concentration on the turbidity level. Figure 2.9(b) shows the percentage of degradation of POMSE for the turbidity under different concentration of the photocatalyst. The results obtained for the three concentrations (25%, 50%, and 100%) are 82% 55.2% and 47.9% respectively. The highest concentration of the POMSE gives the lowest percentage degradation and the lowest concentration of POMSE gives the highest percentage of degradation. These are correlated to the nature of the concentration of organic pollutants where the existence of more organic pollutants can lead to the deactivation of the catalyst. Hence, it can be concluded that at the lowest concentration of POMSE, the highest degradation can be obtained.

The DO value obtained for the different concentration of POMSE was shown in the Figure 2.10 where the value obtained for the three different concentrations (25%, 50%, and 100%) are 6.64 ppm, 5.95 ppm, and 5.6 ppm. The DO value for the lowest concentration is 6.64 ppm which is desirable. This is because of most of the aquatic life need to have DO above 5 ppm according to type of aquatic life. Lower than the said value will cause greater stress to the aquatic life. In order to mimic the ideal environmental systems, the freshwater ideally need around 8 mg/L of DO for optimum growth of the aquatic organism. Therefore, is can be deduced that at 25% concentration of POMSE, the DO obtained is nearest to the ideal condition which indicates that lowest POMSE concentration is safe to be discharged to the environment. This is corresponding to the percentage of color removal and turbidity removal where both indicate the 25% of POMSE concentration as the most efficient (Table 2.3).

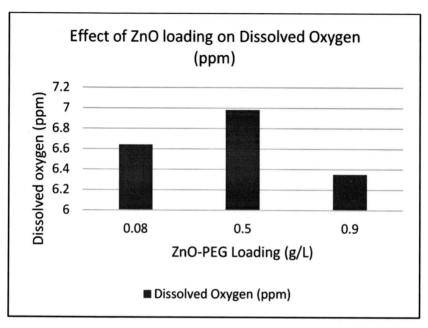

FIGURE 2.8 Value of dissolved oxygen (DO) versus different loading of ZnO-PEG.

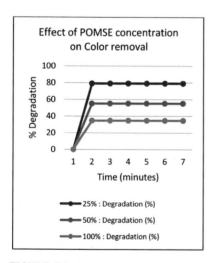

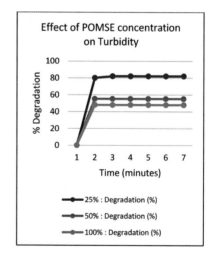

FIGURE 2.9 Percentage of degradation of (a) color removal versus time; (b) turbidity versus time for photocatalysis process in presence of ZnO-PEG (0.015 g/L) under different concentration of POMSE.

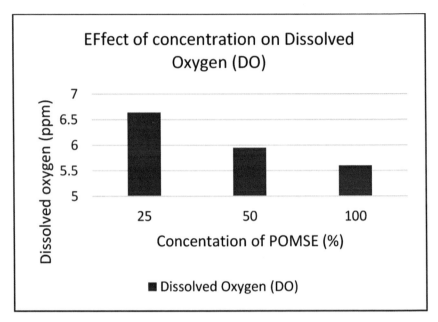

FIGURE 2.10   Value of dissolved oxygen (DO) versus different concentration of POMSE.

TABLE 2.3   Characteristics of POMSE After Photocatalytic Degradation in Presence of 0.08 g/L ZnO-PEG Under Different Initial Concentration of POMSE

POMSE Concentration (%)	POMSE Degradation (%)	Turbidity Reduction (%)	Dissolved Oxygen (ppm)	BOD$_5$ (mg/L)
25	79.0%	82.0	6.64	59
50	55.0%	55.2	5.95	72
100	34.5	47.9	5.60	86

2.5   CONCLUSIONS

The chemical and physical characteristics of ZnO-PEG NPs were successfully characterized by using the XRD and FTIR. Accordingly, the XRD results show that the ZnO-PEG sample was pure and synthesized without any impurities. FTIR results show that the ZnO-PEG sample contains ZnO stretching and deforming vibration. However, the OH compound was detected due to the adsorption of moisture from the surrounding during the synthesis process. The POMSE was successfully treated by using the photocatalytic degradation process in the presence of ZnO-PEG NPs. It was found that the optimum condition for POMSE treatment is under pH 6.5 and 0.08 g/L of ZnO-PEG loading for 25% of the concentration of POMSE. The treated POMSE have the percentage of the degraded color intensity (79%), turbidity (82.5%), DO (6.64 ppm), and BOD (59 mg/L). The analysis of the treated POMSE results shows that this technology have high potential in treating the industrial wastewater with the presence of the ZnO photocatalyst.

ACKNOWLEDGMENT

The authors would like to express their gratefulness to the palm oil plantation and factory located in Johor for providing the industrial POMSE for this work. This research is supported by the Ministry of Education Malaysia under the Fundamental Research Grant Scheme (FRGS) Vot 1616.

KEYWORDS

- **chemical oxygen demand**
- **crude palm oil**
- **fresh fruit bunches**
- **palm oil mill secondary effluent**
- **photocatalysis**
- **zinc oxide nanoparticles**

REFERENCES

1. MPOB, (2009). *Summary of Industry Performance: (2009)*. http://econ.mpob.gov.my/economy/EID_web.htm (accessed on 22 February 2020).
2. Wu, M. K., Windeler, R. S., Steiner, C. K., Bros, T., & Friedlander, S. K., (1993). Controlled synthesis of nanosized particles by aerosol processes. *Aerosol Science and Technology, 19*, 527–548.
3. DOE, (2011). *Annual Report 2011*. Report 983-9119-77-X. Department of Environment, Ministry of Natural Resources and Environment, Malaysia.

4. Ng, K. H., & Cheng, C. K., (2015). A novel photo mineralization of POME over UV-responsive TiO 2 photocatalyst: Kinetics of POME degradation and gaseous product formations. *RSC Advances, 5*(65), 53100–53110.

5. Cheng, C. K., Derahman, M. R., & Khan, M. R., (2015). Evaluation of the photo catalytic degradation of pre-treated palm oil mill effluent (POME) over Pt-loaded titania. *J. Environ. Chem. Eng., 3*(1), 261–270.

6. Chen, D., Sivakumar, M., & Ray, A. K., (2000). Heterogeneous photo catalysis in environmental remediation. *Dev. Chem. Eng. Miner. Process, 8*(5/6), 505–550.

7. Hairom, N. H. H., Mohammad, A. W., & Kadhum, A. A. H., (2015a). Influence of zinc oxide nanoparticles in the nanofiltration of hazardous Congo red dyes. *Chemical Engineering Journal, 260,* 907–915.

8. Hairom, N. H. H., Mohammad, A. W., Ng, L. Y., & Kadhum, A. A. H., (2015b). Utilization of self- synthesized ZnO nanoparticles in MPR for industrial dye wastewater treatment using NF and UF membrane. *Desalination and Water Treatment, 54*(4/5), 944–955.

9. Chiranont, M., (2017). *Optimization of Industrial Dye Wastewater Treatment via Photocatalysis Process in Presence of ZnO-Peg Nanoparticles.* University Tun Hussein Onn Malaysia: Bachelor Degree Thesis.

10. Asfaram, A., Ghaedi, M., Alipanahpour, E., Agarwal, S., & Gupta, V. K., (2015). Application of response surface methodology and dispersive liquid-liquid microextraction by microvolume spectrophotometry method for rapid determination of curcumin in water, wastewater, and food samples. *Food Anal. Methods.*

11. Rauf, M., & Ashraf, S., (2009). Fundamental principles and application of heterogeneous photocatalytic degradation of dyes in solution. *Chemical Engineering Journal, 151,* 10–18.

12. Alhaji, M. H., Sanaullah, K., Lim, S. F., Khan, A., Hipolito, C. N., Abdullah, M. O., & Jamil, T., (2016). Photocatalytic treatment technology for palm oil mill effluent (POME): A review. *Process Safety and Environmental Protection, 102,* 673–686.

13. Idris, M. A., Jami, M. S., & Muyibi, S. A., (2010). Tertiary treatment of biologically treated palm oil mill effluent (POME) using of membrane system: Effect of MWCO and transmembrane pressure. *Int. J. Chem. Environ. Eng., 1*(2), 108e112.

14. Chong, M. N., et al., (2010). Recent developments in photocatalytic water treatment technology: A review. *Water Res., 44*(10), 2997–3027.

15. Gaya, U. I., & Abdullah, A. H., (2008). Heterogeneous photocatalytic degradation of organic contaminants over titanium dioxide: A review of fundamentals, progress, and problems. *J. Photochem. Photobiol. C: Photochem. Rev., 9*(1), 1–12.

16. Saquib, M., (2003). TiO_2-mediated photocatalytic degradation of a triphenyl methane dye (gentian violet) in aqueous suspensions. *Dyes Pigments, 56,* 37–49.

CHAPTER 3

CO_2, D_2O, and Global Warming: How to Stop the Planet from Burning

FRANCISCO TORRENS[1] and GLORIA CASTELLANO[2]

[1]Institute for Molecular Science, University of Valencia, PO Box 22085, E-46071, Valencia, Spain, E-mail: torrens@uv.es

[2]Department of Experimental Sciences and Mathematics, Faculty of Veterinary and Experimental Sciences, Valencia Catholic University Saint Vincent Martyr, Guillem de Castro-94, E-46001 Valencia, Spain

ABSTRACT

It results in significant to be acquainted with who present fake environmental documents. The financial, communal, and following outlays of petroleum reduction results are unprecedented. The cost of nuclear power depends on the following place. The price of every type of power depends on who performs the computations, which results in a following and financial subject. There is merely a single sensibly effective way of amassing electrical energy in an extended period, i.e., forcing storing stations. The entire electrical energy can be created with a couple of kinds of low-C stations, i.e., gas-shot energy stations from which release CO_2 was taken out, and renewable power stations placed in the land of Great Britain or hundreds of kilometers attached to the network through extended-space wires. The electrical energy results in a produced and processed manufactured good, and to employ it as unprocessed stuff to create an additional kind of petroleum results profligate. The mainly significant point results that the scheme does not need additional period and attention of the consumer than he more often than not dedicates to the present power source. This alteration needs a brave rule and determining manufacturing schemes. The genuine transportation difficulty does not result in technical and financial, i.e., it becomes supporting or mental. The train journey results in a smaller amount organized than the bus one.

3.1 INTRODUCTION

Setting the scene: carbon dioxide (CO_2), lauric acid (the fatty constituent of coconut oil), deuterium oxide (D_2O, heavy water), Teflon [poly(tetrafluoroethylene) (PTFE)], global warming (GW), houses with cracks, nuclear energy (NE), renewable energies (REs), interconnected energy, a new transport system, kilometers of love, virtual purchase, and a delayed Apocalypse.

The following question (Q) was raised:

Q1. Do you want to know who show false ecological credentials?

If they are not mitigated in time, the economic, social, and political costs of the depletion of oil will be unheard-of. The price of NE depends on the political position. The cost of all kinds of energy depends on who made the calculations: It is a political as well as economic issue. There is only one reasonably efficient means of storing electricity in the long term: pumping storage plants. All electricity can be produced with two types of low-C generators: gas-fired power plants from whose emission CO_2 was extracted and renewable energy plants located in the territory of GB or hundreds of kilometers connected to it to the net via long-distance cables. Electricity is a manufactured and refined product, and to use it as a raw material to produce another type of fuel is a new extravagance. The most important thing (in terms of public acceptability) is that the system does not require more time and care of the user than he usually devotes to the current supply of energy. Such transformation requires a bold policy and ambitious engineering projects. The real problem of transport is neither technological nor economic: It is political or, rather, psychological. The train trip, which in the popular imagination is associated with low environmental impact, is less efficient than the bus trip. If the car is replaced by the urban and intercity bus, the carbon emission is reduced by 88%. The impact of aviation is negative because the most vulnerable to climate change are the poorest inhabitants of the poorest countries. A reduction of carbon emissions requires that most planes that fly at present remain on the ground.

In earlier publications, it was reported nuclear fusion, the American nuclear cover-up in Spain in 1966 Palomares disaster [1], Manhattan Project, *Atoms for Peace,* nuclear weapons, accidents [2] nuclear science and technology [3]. In the present report, it is reviewed CO2, lauric acid, coconut oil, D2O, Teflon, houses with cracks, NE, REs, interconnected energy, a new transport system, kilometers of love, virtual purchase and a delayed Apocalypse. The aim of this work is to initiate a debate by suggesting a number of questions, which can arise when addressing subjects of CO2, lauric acid, D2O, Teflon, GW, NE, REs. interconnected energy, a new transport system, kilometers of love, virtual purchase and a delayed Apocalypse. It was provided, when possible, answers (A), hypotheses (H) and paradoxes (Pa) on coconut oil, GW, and houses with cracks, NE, REs, and kilometers of love.

3.2 CARBON DIOXIDE (CO$_2$)

Carbon dioxide (CO$_2$) (*cf.* Figure 3.1) is and is not the greenhouse effect (GHE) gas and, what is more, more than one type of GHE exists [4]. The CO$_2$ absorbs energy strongly in two regions of the infrared (IR) part of the electromagnetic spectrum (EMS, as do H$_2$O, CH$_4$, O$_3$). If no CO$_2$ existed in the Earth's atmosphere at all, the Earth's temperature would result in –20°C. Life would be possible but it would not be *hospitable*. A *natural* GHE exists, reducing the loss of long-wavelength (WL) IR heat by radiation, because of *natural* amounts of CO$_2$ in the atmosphere, just under 300 parts per million (ppm). When people talk about *the* GHE, what they are referring to is something extra: the effect on the Earth's temperature caused by CO$_2$ concentrations rising above the *historic* level. Before the Industrial Revolution, CO$_2$ level was 280 ppm, 0.028% of the atmosphere. Daily average CO$_2$ levels (measured at Mauna Loa Observatory in Hawaii) rose from 310 ppm (0.031%) in 1958 to 400 ppm (0.040%), first reached on May 9th, 2013. The CO$_2$ gets into the atmosphere in various ways and, at present, more exists being added to the atmosphere than being removed. *Natural* processes exist that add CO$_2$ to the atmosphere (e.g., wildfires, volcanoes). However, the recent rapid rise in CO$_2$ levels is because of man-made processes, e.g., burning fossil fuels, whether for transportation, providing heat in cold weather or generating power (e.g., coal-fired power stations). In some parts of the world, clearing areas like forests to free up land is an important cause.

FIGURE 3.1   Molecular structure of carbon dioxides.

3.3   LAURIC ACID: THE FATTY CONSTITUENT OF COCONUT OIL

Around 50% of the content of coconut flesh is fatty (FA) lauric (or dodecanoic) acid (*cf.* Figure 3.2) [4]. Eating coconuts is and is not bad for one's heart. The FAs are known to increase cholesterol (CHOL) in the bloodstream, which can lead to blocking of the blood vessels in the heart, producing heart attacks. In addition, lauric acid is known to produce the largest rise in total CHOL of all FAs. However, this is not as bad as it seems because most of the rise is owing to a rise in the *good* CHOL. Medics classified CHOL into two types: high-density lipoprotein (HDL) and low-density lipoprotein (LDL), depending on the density of the lipoproteins. The LDLs are bad for people, as they greatly increase the chance of getting arterial blockages, whereas HDLs are not such a problem. As a result, medics believe that lauric acid lowers the ratio of total CHOL to HDL CHOL more than any other FA, either saturated or unsaturated. Eating coconuts may decrease the risk of getting heart disease, although that must to be proven. It comprises 6.2% of the total fat in human breast milk, 2.9% of the total fat in cow's milk, and 3.1% of total fat in goat's milk. Lauric acid is known to the pharmaceutical industry for its excellent antimicrobial properties, and the monoglyceride derivative of lauric acid, monolaurin, is known to present even more potent antimicrobial properties *vs.* lipid-coated ribonucleic (RNA) and deoxyribonucleic (DNA) acid viruses, numerous pathogenic Gram$^+$ bacteria and a number of pathogenic protozoa. As well as foodstuffs, coconut oil can be used as a feedstock for biodiesel fuel. A number of tropical island countries are using coconut oil as a fuel source to power Lorries, cars, and buses, and for electrical generators. Before electrical lightning became commonplace, coconut oil was used for lighting in India and was exported (*Cochin oil*). Coconut oil can be used as a skin moisturizer, as engine lubricant and transformer oil, and acids derived from coconut oil can be used as a weed-killer. Coconut oil is used by movie theater chains to pop popcorn!

When one eats lauric acid, much of it gets converted into monolaurin (glyceryl laurate, *cf.* Figure 3.3), which may help protect vs. bacterial infection. Because of this, monolaurin is commonly used in deodorants. Finally, monolaurin gets converted into HDL CHOL.

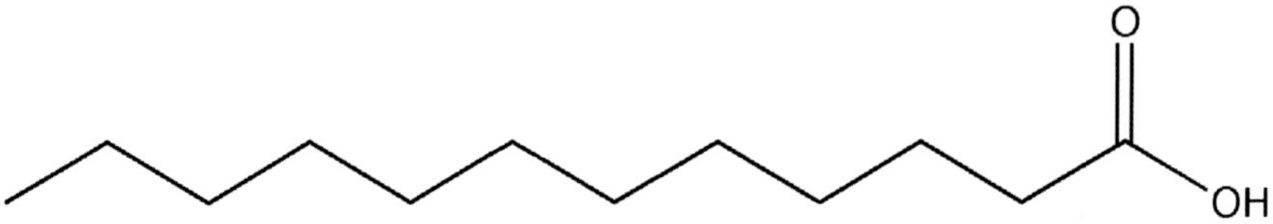

FIGURE 3.2 Molecular structure of lauric acid.

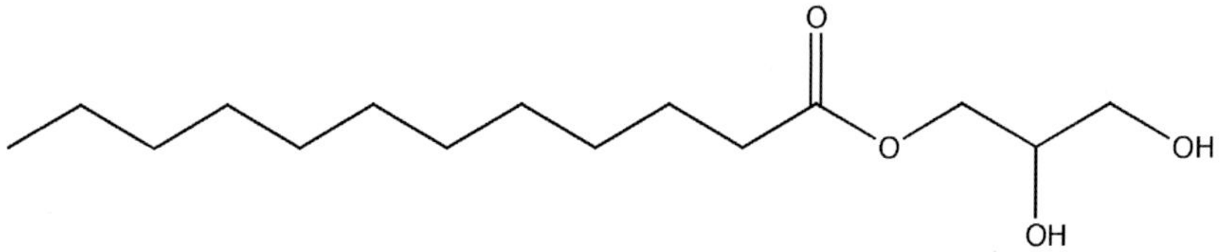

FIGURE 3.3 Monolaurin, the monoglyceride formed joining one lauric acid molecule to glycerol.

Because lauric acid is relatively cheap, it presents a long shelf-life, is non-toxic and safe to handle, it is usually used for the production of soaps and cosmetics. The lauric acid is extracted, then neutralized with NaOH to give Na+ laurate, which is a soap, which can be further processed to give sodium lauryl sulfate (SLS, *cf.* Figure 3.4), a detergent. The SLS [sodium dodecyl sulfate (SDS)] is the main component of most soap-based products, and if one were to look in his bathroom, he is guaranteed to find at least one product containing it (e.g., shampoo, toothpaste). In industry, it is found in products (e.g., engine degreasers, carpet cleaners). The SLS is inexpensive and an excellent foaming agent. It presents a high pH as it is an alkali substance and shows the appearance of a white powder. The SLS is a by-product of palm oil, which is extracted from palm trees, which are mainly grown in Indonesia and Malaysia, and accounts for a large percentage of income for much of the local population. However, in order to make way for palm oil plantations, mass deforestation occurred in the countries. Deforestation on such a large scale presents severe implications for the climate of the whole planet. When cutting down ancient rainforests, great quantities of C are released into the atmosphere from peat bogs, adding hugely to GW. The loss of trees that carry out photosynthesis (the process of converting CO$_2$ to O$_2$) greatly adds to GW problem, since the palm oil trees that replace them are not as large, as densely packed, or as efficient at photosynthesis as those in the rainforest. Many environmental groups claimed that the deforestation being carried out in order to make room for plantations is actually more harmful to the Earth's climate than the benefits that could be gained *via* palm oil as a biofuel.

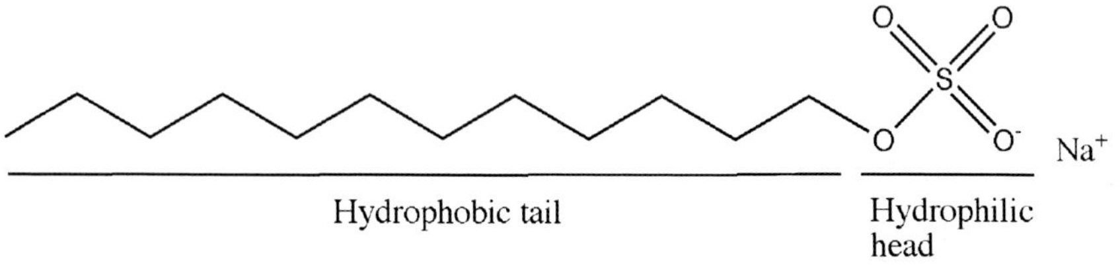

Hydrophobic tail	Hydrophilic head

FIGURE 3.4   Molecular structure of sodium lauryl sulfate (SLS).

3.4 DEUTERIUM OXIDE (HEAVY WATER)

For most practical purposes, substituting one isotope for another in a compound presents no effect on bulk physical properties [5]. In the case of such a light molecule as water, the difference between H_2O and D_2O leads to significantly different melting points (0.00 and 3.81°C, respectively) and boiling points (100.00 and 101.42°C); G. N. Lewis and his student, Ronald T. MacDonald, were able to obtain pure D2O (*cf.* Figure 3.5) by fractional distillation under reduced pressure in 1933. Molecular dimensions of the isolated molecules are slightly different; for an isolated H_2O molecule, O-H is 0.9724Å and H-O-H is 104.50°, whilst for an isolated D_2O molecule, O-D is 0.9687Å and D-O-D is 104.35°. In the condensed liquid phase, the O-H bond in H2O is 1.01Å, whilst the O-D bond in D_2O is 0.98Å, shorter by 0.03Å; on the other hand, the intermolecular bond in D_2O is longer by 0.07Å. One significant difference between H_2O and D_2O is that D_2O can act as a moderator in a nuclear reactor to produce neutrons of the right velocity to induce nuclear fission in [235]U, which, in the late 1930s and early years of World War II (WW2), led to competition between the Germans and French for supplies of D_2O and, subsequently, to attacks on the Rjukan plant in Norway that produced it; the film *Heroes of Telemark* was loosely based upon these events. Since WW2, chemists made use of the different H and D nuclear properties differently: a compound with labile N-H or O-H groups will exchange the Hs on shaking with D2O, so the signal because of the Hs in 1H-nuclear magnetic resonance (NMR) spectrum of the compound will disappear, facilitating spectrum assignment.

$$R - O - H + D_2O \leftrightarrow R - O - D + D - O - H$$

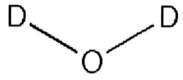

FIGURE 3.5 Molecular structure of deuterium oxide (heavy water).

3.5 TEFLON, POLY(TETRAFLUOROETHENE) (PTFE)

Teflon was discovered in 1938 by U.S. chemist Roy Plunkett; trying to make a chlorofluorocarbon (CFC) refrigerant, he heated tetrafluoroethylene (*cf.* Figure 3.6a) in a Fe container under pressure, obtaining a white solid [5]. It transpired that Fe catalyzed the polymerization, leading to PTFE (*cf.* Figure 3.6b). The strong C-F and C-C bonds cause it to be chemically inert to other chemicals, making it of almost immediate use in the Manhattan project for the construction of the first nuclear bomb. The isotopes ^{235}U and ^{238}U were separated by gaseous diffusion of volatile and reactive UF_6; Teflon was vital for making unreactive and leak-proof valves and seals in the plant. Today, Teflon-coated magnetic stirrer bars are widely used in laboratories. Teflon presents an extremely low coefficient of friction and is widely used for a range of nonstick applications [Teflon-coated, nonstick cookware, coatings for machinery parts (gears, bearings), windscreen wiper blades]. Teflon taps in the burettes prevent sticking and require no lubrication. Teflon tape is used as a thread-sealing tape by plumbers, whilst Teflon coatings are used on some armor-piercing bullets (though it is not the Teflon that gives them the armor-piercing property).

3.6 FAUSTUS' PACT

Monbiot proposed the following question and hypothesis on GW [6]:

Q1. How to stop the planet from burning?
H1. Faustus' pact: The use that people do of fossil fuels.

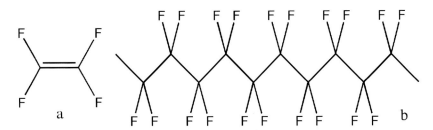

FIGURE 3.6   Molecular structures: a) tetrafluoroethene and b) poly(tetrafluoroethene).

He raised the following questions on rejecting the explanation of planetary GW:

Q2. Does the atmosphere contain CO_2?
Q3. Does atmospheric CO_2 increase the mean global temperature?
Q4. Will such an effect rise by the addition of more CO_2?
Q5. Have human activities produced a net CO_2 emission?

He proposed the following additional questions and answer:

Q6. Then, what will occur if mean global temperatures rise more than 6°C?
Q7. Is to stop climate change possible?
Q8. [7]. Is it too late [7]?
A8. No, people has a short time.

3.7   THE INDUSTRY OF NEGATION

He proposed the following questions, answers, and hypotheses on the industry of negation:

Q1. What are C emissions (CEs)?
Q2. Why are CEs a problem?
Q3. Why have people taken so long time to act?
A3. The problem is not that people do not hear about GW but that they do not want to know.
H1. Untrue H. Pollution is a heaven's gift.
Q4. [8]. GW [8]?
Q5. [8]. Why do people accept these lies like if they wear blinkers?
Q6. [9]. GW [9]?
Q7. Where did data come?

H2. (Brown and Williamson). Doubt is our object as best way of competing with *dataset* in public's mind.

Q8. (BBC). What does the future provide on climate change (CC)?

Q9. (US Senator Inhofe). With all hysteria, fears, false science, would not man-induced GW be greatest story?

Q10. How did he know it?

A10. Because he had spoken to the main climatologists of the country.

Q11. (King). Well, in which moment will the ice layer of Greenland begin melting?

Q12. (King). Which CO_2 level is it essential not to exceed not to have to verify such theory?

H3. Never disturbances occurred to ask for austerity.

3.8 A RATIO OF FREEDOM

He proposed the following hypotheses and questions on a ratio of freedom:

H1. *Contraction/convergence Principle*: Rationing effect would be global-CEs recession.

H2. C-rationing-created market will automatically encourage low-CE technologies (e.g., public transport, REs).

H3. C-rationing programme will not be fair if it does not go with poorest' housing programme.

H4. The more energy C quota they allow using, the more politically acceptable programme will be.

H5. Kyoto protocol/EU emissions trading system (ETS): the dirtier one is, the more right he has.

H6. [10]. *The Skeptical Environmentalist* [10].

H7. [10]. The economic recession will be higher than the losses that CC could cause.

Q1. Is it possible to assign an economical price to human life?

Q2. Or to an ecosystem?

Q3. Or to climate?

H8. (British Government). The *social cost* of C emissions is 35–140£ per ton.

Q4. What the devil can that figure mean?

Q5. Does the British Government really believe that prize to Amazon could be put?

Q6. Or to Bangladesh?

H9. [11]. *Perverse Subsidies* [11].

Q7. [11]. How can tax dollars undercut the environment and the economy?

Q8. Why do governments find money to destroy biosphere and so many difficulties to save it?

Q9. When will the global supply of oil reach a maximum?

H10. [12]. *Peaking Oil of World Oil Production* [12].

H11. [12]. If not mitigated, economic/social/political costs will be unheard-of.

3.9   HOUSES WITH CRACKS

He proposed the following questions, hypotheses, and paradoxes on houses with cracks:

Q1. (Marlowe). Who was the one that built so badly the house, with shovel and pick [13]?

H1. [14]. *The Coal Question* [14].

Q2. [14]. Which is Nation's progress faced with probable exhaustion of our coal mines?

Q3. [14]. Can Britain survive?

Q4. (Herring). Does energy efficiency save energy?

Q5. Why companies, in their effort to reduce costs, have not saved the planet?

H2. Untrue H. Energy efficiency is equivalent to energy reduction.

Pa1. Khazzoom-Brookes' postulate: In a free market, energy efficiency increases energy use [15].

Pa2. Bounce effect: One living in a well-isolated house will be tempted to increase the temperature.

H3. Lacking government policy, energy efficiency is not only a waste of time but antiproductive.

Pa3. Regulation increases the sum of human freedom.

3.10   THE TURNED-ON LIGHTS

He proposed the following questions, answers, and hypothesis on NE:

Q1. How do you judge the danger of CC in comparison with the risk of a nuclear war?

A1. The former is real and occurring; the latter is like (perhaps more) devastating but less safe.

Q2. Why deceit is frequent in nuclear industry?

A2. Because it is cheaper to manipulate badly; nuclear managers have the perfect excuse (safety).

H1. Elementary environmental principle: Do not stir anything again until you order what you stirred.

Q3. Why Government is so generous in comparison with, e.g., wind power (WP)?

A3. NE served as armament deterrent; the more expensive a project, the more powerful pressure group wants it to be approved.

Q4. Well, how much does NE cost?

A4. The price of NE depends on the political position.

He proposed the following three important questions and answers.

Q5. Is there U (the main nuclear fuel) enough to sustain such industry?

A5. It depends on a number of unforeseeable factors.

Q6. How much CO_2 is saved *via* NE?

A6. The worse be the quality of a U ore, the greater energy is required to process it.

Q7. Can NE be distributed quickly enough to carry out our objective?

A7. If governments decide CC is urgent, they can transform economy in the twinkling of an eye.

He proposed the following additional questions:

Q8. How much electricity can REs supply?

Q9. At which cost?

3.11 HOW MUCH ENERGY CAN RENEWABLE ENERGIES (RES) PROVIDE?

He proposed questions, answer, and hypotheses on how much energy REs can provide:

Q1. How much energy can REs provide?

Q2. Have people enough RE power plants?

Q3. In addition, plans beyond 20–30% of total energy?

A3. It seems nobody plan it.

Q4. Is such evaluation reliable?

Q5. Who did he make the calculations?

H1. Cost of all kinds of energy depends on who made calculations: It is a political/economic issue.

H2. Only one reasonably efficient means of storing electricity in the long term: pumping storage plants.

Q6. Can more pumping storage plants be built?

He proposed additional hypotheses on REs:

H3. In the UK, enough RE exists to cover comfortably an average of 50% of electricity consumption.

H4. The net, and reliability of electricity it distributes, resists if that 50% of supply come from REs.

H5. The C costs of generating it are significantly lower than the C savings.

H6. Price per kW·h does not double that British Government offered for WP replacing 20% electricity.

H7. Global H: The 50% goes into the field of viability.

H8. The other 50% is provided by thermal power plants, which CEs were captured and stored.

H9. [16]. *When the Rivers Run Dry* [16].

H10. Growing wood for heating involves fewer moral hazards than biodiesel: woody plants can be grown in poorer lands; growing firewood to burn is more efficient method to reduce CO_2; wood-production value per ha is less.

3.12   THE INTERCONNECTED ENERGY

He proposed the following questions, answers, and hypotheses on the interconnected energy:

Q1. How can electric energy be generated?

A1. By *microgeneration* (*interconnected energy*).

Q2. However, would it resist an examination?

Q3. Does such technology result in a *more or less* reasonable price?

H1. Two main problems in the UK: There is not enough sun and it shines when it should not.

H2. [17]. *Half Gone* [17].

Q4. How much RE has the country?

H3. Problem: The supply of photovoltaic (PV) electricity is very little adjusted to the demand.

H4. Untrue H. A home with solar panels can sell more electricity to the net than the one that buys.

H5. Problem: Microaerogenerators only produce wind electricity when wind is strong and constant.

Q5. (Martin). Can we harvest useful wind energy from the roofs of our buildings?

H6. [18]. *Wind Energy for the Rest of Us* [18].

H7. *Domestic heat/energy combination*: Use tiny power plant to produce heat and electricity.

H8. Electricity is manufactured/refined product; use as raw material for another fuel is extravagant.

H9. (Andrews). To produce H_2 in homes from electricity *via* semi-electrolysis.

Q6. Will a microgeneration system work without repeated blackouts?

A6. Probably.

H10. Dismantling the natural net would be a mistake: It would be to fall into the aesthetic fallacy.

H11. The most important for acceptability is that the system does not require time and user care.

H12. Such transformation requires a bold policy and ambitious engineering projects.

3.13   A NEW TRANSPORT SYSTEM

He proposed the following hypotheses and paradox on a new transport system:

H1. The real problem of transport is not technological/economic: It is political/psychological.

Pa1. Mass movement: System that makes people cattle but guaranteeing them autonomy illusion.

H2. The richer people are, the more they can travel but the fewer places they have to venture.

He proposed the following conclusions:

C1. The train trip, which is associated with low environmental impact, is less efficient than the bus.

C2. If the car is replaced by the urban and intercity bus, CE is reduced by 88%.

He proposed the following additional hypotheses, questions, and answer:

H3. The problem is that people hate the bus: The trip is an exasperating and humiliating experience.

H4. Coaches became resource of those who cannot afford other means (poverty/marginalization).

H5. Politically, coach travel should be faster, more relaxing, and more reliable than car travel.

H6. (Storkey). Move the bus stations from the urban center to the motorway accesses.

Q1. (Collins). Why did you reject strategic-authority recommendation to create high-quality-coaches net on ring roads?

H7. As measure to contain CC, Storkey's H results complementing it with policies: rationing CEs; road space people use.

H8. Biofuel problem: people have limited agricultural land/water; impact condemns millions to hunger rising food price.

Q2. (British Government). Does the import ban violate international trade laws?

A2. (E_4Tech). Imperative environmental criteria increase international-law-infringement risk.

H9. Cars, like houses, can be equipped with H_2 fuel cells.

H10. [19]. *Natural Capitalism* [19].

H11. Obstacles: H_2 cannot be purchased at service stations; it has low energy density in terms of volume/needs 10× larger high-pressure reservoir.

H12. Electric-vehicle (EV) problem: autonomy limited by batteries capacity (spent at 170–500 km, take hours to recharge).

H13. [20]. *Car Sick* [20].

H14. [20]. *40:40:20 rule*: Of current car trips, 40% can be made by bicycle/foot/public transport; 40%, if facilities were improved; 20% are irreplaceable.

H15. [20]. One of the main problems is that people do not know the existing services.

H16. It is necessary to improve significantly the urban bus services.

3.14 KILOMETERS OF LOVE

He proposed the following hypotheses and questions on kilometers of love.

H1. People of goodwill are as capable of destroying the biosphere as the executives of Exxon.

H2. Untrue H. Buddhist H: It does not matter what you do if you do it with love.

H3 To think, dress, and decorate the house is of no use unless people behave like ethical people.

Q1 Who can be surprised, therefore, to prove that *ethical* people deny the impact of flying?

H4 Subsonic-air-travel impact: great covered distance; gases/particles release, giving a 2.7× GHE.

H5 Supersonic aircraft fly *via* stratosphere and produced water vapor causes a 5.4× GHE impact.

H6 *Prediction/forecasting* focus: The more space increased to respond to demand prediction, this rises until occupying it and it is needed to create more space to make square new predictions.

H7 Untrue H. British Ministry of Transport: Airports-capacity enlargement is a *social measure*.

H8 Aviation impact is negative because most CC vulnerable is poorest countries' poorest inhabitants.

H9 Social classes cannot afford holiday abroad.

H10 Two ways: a large-scale rise in energy efficiency of planes; a new fuel.

H11 A reduction of CEs requires that most planes that fly at present remain on the ground.

Q2 Are there other ways to cover the same distance at the speed that people is accustomed?

Q3 However, should a series of transcontinental high-speed railways be done?

H12 (Kemp). The consumption of energy rises spectacularly at speed above 200 km·h^{-1}.

H13 To admit that progress is now based on exercising less opportunities.

3.15 VIRTUAL PURCHASE

He proposed the following questions, answer, and hypotheses on retail trade:

Q1. [21]. *How to Eat* [21]?

Q2. [21]. Why not to thank for livings in jet-transport age/generalized gastronomic-goods importation?

Q3. (Martin). Can we harvest useful wind energy from the roofs of our buildings?

Q4. How could people that go shopping outside city be persuaded to travel in other means?

A4. Delivery.

H1. Shops placed outside cities are substituted by stores to which only suppliers/enterprises personnel come.

H2. Pleasant packing is not necessary.

He proposed the following hypothesis on the manufacture of cement:

H3. A ton of cement produces *more or less* a ton of CO_2.

3.16 A DELAYED APOCALYPSE

He proposed the following hypotheses, questions, and answers on a delayed apocalypse:

H1. Faith in miracles turns into a perfect excuse for inaction.

H2. Miracles: future technologies; new technology will allow removing CO_2 after it were emitted to air or cool artificially planet; oil depletion; buy solution.

H3. Objections: Projects accurate accounting to counteract C is impossible; to arrive to big C cutting, all must limit CEs.

H4. The necessary reduction of CEs is, although difficult, technical, and economically possible.

Q1. Is the necessary reduction of CEs politically possible?

A1. It does not depend on me: it depends on you.

Q2. Why does it result easier to persuade people to protest about other causes than CC?

A2. Other causes do people something, while in GW they are who do them it.

Q3. How have people arrived to that?

A3. Sustained global growth; indebtedness; Internet creates a false impression of action.

3.17 FINAL REMARKS

From the presents results and discussion, the following final remarks can be drawn:

1. It is important to know who show false ecological credentials.
2. If they are not mitigated in time, the economic, social, and political costs of the depletion of oil will be unheard-of.
3. The price of NE depends on the political position.
4. The cost of all kinds of energy depends on who made the calculations: It is a political as well as economic issue.
5. There is only one reasonably efficient means of storing electricity in the long term: pumping storage plants.
6. All electricity can be produced with two types of low-C generators: gas-fired power plants from whose emission CO_2 was extracted and renewable energy plants located in the territory of Great Britain or hundreds of kilometers connected to it to the net *via* long-distance cables.
7. Electricity is a manufactured and refined product, and to use it as a raw material to produce another type of fuel is a new extravagance.
8. The most important thing (in terms of public acceptability) is that the system does not require more time and care of the user than he usually devotes to the current supply of energy.
9. Such transformation requires a bold policy and ambitious engineering projects.
10. The real problem of transport is neither technological nor economic: It is political or, rather, psychological.
11. The train trip, which in the popular imagination is associated with low environmental impact, is less efficient than the bus trip.
12. If the car is replaced by the urban and intercity bus, the carbon emission is reduced by 88%.
13. The impact of aviation is negative because the most vulnerable to climate change are the poorest inhabitants of the poorest countries.
14. A reduction of carbon emissions requires that most planes that fly at present remain on the ground.

ACKNOWLEDGMENTS

The authors thank the support from Generalitat Valenciana (Project No. PROMETEO/2016/094) and Universidad Católica de Valencia *San Vicente Mártir* (Project No. 2019-217-001UCV).

KEYWORDS

- **carbon dioxide**
- **delayed apocalypse**
- **deuterium oxide**
- **Faustus' pact**
- **interconnected energy**
- **lauric acid**
- **love kilometer**
- **negation industry**
- **nuclear energy**
- **poly(tetrafluoroethylene)**
- **renewable energy**
- **Teflon**
- **transport system**
- **virtual purchase**

REFERENCES

1. Torrens, F., & Castellano, G., (2019). Nuclear fusion and the American nuclear cover-up in Spain: Palo mare's disaster (1966). In: Haghi, R., & Torrens, F., (eds.), *Engineering Technology and Industrial Chemistry with Applications* (pp. 297–308). Apple Academic-CRC: Waretown, NJ.
2. Torrens, F., & Castellano, G. (2020). Manhattan project, atoms for peace, nuclear weapons, and accidents. In: Pogliani, L., Torrens, F., & Haghi, A. K., (eds.), *Molecular Chemistry and Biomolecular Engineering: Integrating Theory and Research with Practice* (pp. 215–233). Apple Academic-CRC: Waretown, NJ.

3. Torrens, F., & Castellano, G. Nuclear science and technology. In: Pogliani, L., Ameta, S. C., & Haghi, A. K., (eds.), *Chemistry and Industrial Techniques for Chemical Engineers*. Apple Academic-CRC: Waretown, NJ, in press.

4. May, P., & Cotton, S., (2015). *Molecules That Amaze Us*. CRC: Boca Raton. FL.

5. Cotton, S., (2012). *Every Molecule Tells a Story*. CRC, Boca Raton (FL).

6. Monbiot, G., (2009). *Heat: How to Stop the Planet from Burning*. Penguin: London, UK.

7. Lovelock, J., (2006). *The Earth is About to Catch a Morbid Fever Than may Last as Long as 100,000 Years* (p. 1). The Independent.

8. Hitchens, P., (2001). *Global Warming? It's Hot Air and Hypocrisy* (p. 1). The Mail on Sunday.

9. Bellamy, D., (2004). *Global Warming? What a Load of Poppycock!* (p. 1). Daily Mail.

10. Lomborg, B., (2001). *The Skeptical Environmentalist*. Cambridge University: Cambridge, UK.

11. Myers, N., & Kent, J., (2001). *Perverse Subsidies: How Tax Dollars Can Undercut the Environment and the Economy*. Island: Washington, DC.

12. Hirsch, R. L., Bezdek, R., & Wendling, R., (2005). *Peaking Oil of World Oil Production: Impacts, Mitigation, and Risk Management*. US Department of Energy: Washington, DC.

13. Marlowe, C., (2005). *Doctor Faustus*. Norton: New York, NY.

14. Jevons, W. S., (1865). *The Coal Question: An Inquiry Concerning the Progress of the Nation, and the Probable Exhaustion of Our Coal Mines*. Macmillan: London, UK.

15. Khazzoom, J. D., (1980). Economic implications of mandated efficiency standards for household appliances. *Energy J., 1*, 21–39.

16. Pearce, F., (2006). *When the Rivers Run Dry*. Eden Project: London, UK.

17. Leggett, J., (2005). *Half Gone: Oil, Gas, Hot Air, and the Global Energy Crisis*. Portobello: London, UK.

18. Gipe, P., (2016). *Wind Energy for the Rest of Us: A Comprehensive Guide to Wind Power and How to Use It*. Wind-Works.org: Bakersfield, CA.

19. Hawken, P., Lovins, A. B., & Lovins, L. H., (1999). *Natural Capitalism: The Next Industrial Revolution*. Earthscan: London, UK.

20. Sloman, L., (2006). *Car Sick: Solutions for Our Car-Addicted Culture*. Green: Totnes, Devon, UK.

21. Lawson, N., (2002). *How to Eat: The Pleasures and Principles of Good Food*. Wiley: Hoboken, NJ.

CHAPTER 4

World of Conductive Two-Dimensional Metal-Organic Frameworks

FRANCISCO TORRENS[1] and GLORIA CASTELLANO[2]

[1]*Institute for Molecular Science, University of Valencia, PO Box 22085, E-46071, Valencia, Spain, E-mail: torrens@uv.es*

[2]*Department of Experimental Sciences and Mathematics, Faculty of Veterinary and Experimental Sciences, Valencia Catholic University Saint Vincent Martyr, Guillem de Castro-94, E-46001 Valencia, Spain*

ABSTRACT

A way of transforming matter is manufacturing nanomaterials (NMs)/ devices or composites/layers. More is different. Layered metal-organic (MO) frameworks are modular solids with a width of a few atomic coatings. What are they used for? They exhibit properties distinct from the bulk. Layered materials are obtained and characterized as thin-layered nanostructures. Layered MO frameworks are interesting for materials and applications, owing to their atomically thin-layered structure, large surface chemical reactivity, high surface/volume quotient, and large chemical-adsorption ability. Their structural, electronic, optical, and magnetic properties are utilized for learning. The interlayer distance is the greatest for MoS_2 because of its layered structure, in which a plane of Mo atoms is inserted in other planes of S^{2-}. The three strata form a monolayer of MoS_2. The table of periodic properties of the elements for some layered materials is analyzed: A steadiness results in interlayer distances for B-N in period 2. Model principles and structure-property relationships are derived: A series of linear models of the interatomic distance between coatings, *vs.* atomic number of layered materials, results in that the distance raises with atomic number. Non-linear equations improve correlations. Interlayer distance for S diverges from fits because of the three-stratum monolayer of MoS_2.

4.1 INTRODUCTION

Here is an example of MO framework (MOF, *cf.* Figure 4.1). Layered-nanomaterials (NMs) questions follow [1]:

Q1. (Feynman). What could we do with layered structures with just the right layers?

Q2. (Feynman). What would the properties of materials be if we could really arrange the atoms the way we want them?

What can be done in the regard of classification for MOFs?

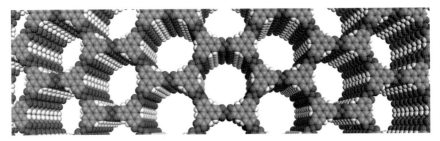

FIGURE 4.1   Conductive MOF with hexagonal pore apertures formed with triphenylene (HXTP) ligands. *Source:* Mirica [73].

The MOFs combine the properties of their inorganic and organic parts (*cf.* Table 4.1). In condensed matter physics (CMP), *more is different* [2]. Geim group reported bulk-graphite exfoliation into graphene (GR) [3], which paved the way to other two-dimensional (2D) crystals [4]. In this laboratory, Coronado group published superconductivity (SC)-magnetism co-existence by chemical design [5]. They imagined the magnetic reversal of isolated and organized molecular-based nanoparticles (NPs), *via* magnetic force microscopy (MFM) [6]. They switched magnetic vortex core in an only NP [7]. To induce quantum confinement effects drives current interest in one-monolayer (ML)-thick 2D NMs [8].

In earlier publications, it was informed the effects of type, size, and elliptical deformation on molecular polarizabilities of model single-wall carbon nanotubes (SWNTs) from atomic increments [9–12], SWNTs periodic properties and table based on the chiral vector [13, 14], calculations on SWNTs solvents, co-solvents, *cyclo* pyranoses [15–18] and organic-solvent dispersions [19, 20], packing effect on cluster nature of SWNTs solvation features [21], information entropy analysis [22], cluster origin of SWNTs transfer phenomena [23], asymptotic analysis of coagulation-fragmentation equations of SWNT clusters [24], properties of fullerite and symmetric C-forms, similarity laws [25], fullerite crystal thermodynamic characteristics, law of corresponding states [26], cluster nature of nanohorns (SWNHs) solvent features [27], SWNTs (co-)solvent selection, *best* solvents, acids, superacids, host-guest inclusion complexes [28], C/BC$_2$N/BN fullerenes/ SWNTs/nanocones (SWNCs)/SWNHs/buds (SWNBs)/GRs cluster solvation models in organic solvents [29–35], elementary polarizability of Sc/ *fullerene*/GR aggregates, di/GR-cation interactions [36] and conductive layered MOFs [37–40]. Conductive 2D MOFs were examined *via* instances from the Mirica group as examples.

TABLE 4.1 Metal-Organic Frameworks Combine the Properties of Their Inorganic and Organic Parts

By its inorganic part	By its organic part
Extended structure	Synthetic versatility
High crystallinity	Easy processability
Diversity	

4.2 UNIQUE CHARACTERISTICS AND CHALLENGES OF ULTRA-THIN 2D NMS

The NMs present a high surface-to-volume ratio and are more active than their micro and macro counterparts. Why 2D NMs? Their properties follow: (1) atomically thick, (2) control over the thickness, and (3) free of defects (pinhole-free device). Atomic sheets of 2D are atomically thin, layered crystalline solids with the defining characteristics of intralayer covalent bonding and interlayer van der Waals (VDW) bonding. An MOF is an ordered nanoporous three-dimensional (3D) coordination polymer composed of an inorganic part, an ion, or cluster, co-ordinated by organic *linkers*. It is a co-ordination net with organic ligands containing potential voids, or a co-ordination net being co-ordination compounds extending in at least one dimension *via* repeating co-ordination entities. The MOFs are net materials comprised of organic ligands connected by metal ion clusters into multidimensional structures, which usually present permanent porosity.

The unique characteristics of ultra-thin 2D NMs follow: (1) the electron confinement in 2D of ultra-thin 2D NMs without interlayer interactions, especially single-layer nanosheets, enables greatly compelling electronic properties compared to other NMs, rendering them appealing candidates for fundamental CMP study, electronic, optical, and magnetic nanodevice (ND) applications, (2) the atomic thickness offers them maximum mechanical flexibility and optical transparency, making them promising for the fabrication of highly flexible and transparent electronic/optical-electronic (*optoelectronic*) NDs, (3) the large lateral size and ultra-thin thickness endow them with ultra-high specific surface area, making them highly favorable for surface-active applications (chemical reactivity, adsorption, catalysis, etc.), (4) magnetic MOFs, (5) biological MOFs (bio-MOFs), (6) supramolecularly organized MOFs (SOFs), etc.

The MOFs present numerous fascinating applications associated with their important porosity and chemical versatility, e.g., gases separation, sensors, catalysts, biomedicine use, etc. Beyond the direct applications, MOFs serve for molecular sponges, which allow incorporating inside different molecules in an orderly way, facilitating the structural determination of the molecules that, otherwise, would be impossible. However, MOFs are considered to be inert for the fabrication of electronic NDs, owing to their poor electrical conductivity and difficult film-forming ability. The construction of conductive or semiconductive 2D MOFs is one of the promising possible solutions for MOFs integration into electronic NDs.

4.3 ELECTRONICS AND OPTOELECTONICS OF TWO-DIMENSIONAL (2D) TMDS

GR remarkable properties present renewed interest in inorganic, 2D NMs with unique electronic and optical attributes. Transition metal dichalcogenides (TMDs) are layered NMs with strong in-plane bonding and weak out-of-plane interactions, enabling exfoliation into 2D layers of single unit-cell thickness. Although TMDs were studied for decades, recent advances in NMs characterization and ND fabrication opened up opportunities for 2D layers of thin TMDs in nanoelectronics and optoelectronics. The TMDs (e.g., MoS_2, $MoSe_2$, WS_2, WSe_2) present sizable bandgaps that change from indirect to direct in single layers, allowing applications [e.g., transfer variators (*transistors*), photodetectors, electroluminescent devices]. Strano group reviewed the historical development of TMDs, methods for preparing atomically thin layers, electronic, and optical properties, and prospects for future advances in electronics and optoelectronics [41].

Fukagawa group proposed novel Pt complexes and a suitable host/dopant combination to realize red phosphorescent organic light-emitting diodes (LEDs) (OLEDs) (PHOLEDs), whose low power consumption and high stability were shown [42]. They showed that the host/dopant combination is crucial for ensuring the efficiency and stability of Pt complexes. The efficient energy transfer from bis(benzo[h]quinolin-10-olato-kN,kO) beryllium^{II} (Bebq$_2$) to the Pt complexes was confirmed *via* the host- and/or electron-transporting layer (ETL)-dependent device characteristics. The optimized PHOLED produced *via* Bebq$_2$/Pt complex combination exhibited color saturation with Commission Internationale de l'Eclairage (CIE) coordinates of (0.66, 0.34), low driving voltage, high efficiency, and high stability. A maximum external quantum efficiency (EQE) over 19% was obtained with a maximum power efficiency of 30lm·W^{-1}, which is comparable to the highest values reported for red PHOLEDs *via* Ir complexes. The estimated half-life of the optimized PHOLED was 10000 h with an initial luminance of 1000 cd·m^{-2}. They found that Pt complexes could act as efficient dopants with efficiencies close to Ir complexes, depending on the host material, which finding could expand the range of design possibilities of novel phosphorescent dopant materials for efficient and stable PHOLEDs.

4.4 ULTRA-THIN TWO-DIMENSIONAL (2D) NANOMATERIALS (NMS)

GR's success showed that it is possible to create stable, ML, and few-atom-thick layers of VDW NMs, and that NMs exhibit fascinating and technologically useful properties. Goldberger group reviewed 2D-NMs state-of-the-art beyond GR [43]. They outlined the different chemical classes of 2D NMs, and discussed a number of strategies to prepare ML, few-, and multilayer assembly NMs in solution, on substrates and wafer scale. They presented an experimental guide for identifying and characterizing ML NMs, and outlining emerging techniques that yielded local and global information. They described the differences that occur in the electronic structure between the bulk and ML, and discussed a number of methods of tuning their electronic properties by manipulating the surface. They highlighted ML properties and advantages, few-, and many-layer 2D

NMs in field-effect transistors (FETs), spin-, and valley-tronics, thermo-electrics, and topological insulators, etc. GR 2D atomic sheets represent a new NMs class with potential applications in electronics. However, exploiting the intrinsic characteristics of GR devices was problematic, because of impurities and disorder in the surrounding dielectric and GR/dielectric interfaces. Recent advancements in fabricating GR heterostructures, alternately layering GR with crystalline hexagonal boron nitride (BN-*h*), its insulating isomorph, led to an order of magnitude improvement in GR device quality. Shepard group reviewed recent developments in GR devices *via* BN dielectrics [44]. They examined the systems' FET characteristics at high bias. They discussed existing NM-synthesis and fabrication challenges, and the potential of GR/BN heterostructures for electronic applications.

The compelling demand for higher performance and lower power consumption, in electronic systems, is the main driving force of the electronics industry's quest for NDs and/or NM-based architectures. Bonaccorso group reviewed electronic NDs based on 2D NMs, outlining potential as a technological option beyond scaled complementary metal-oxide-semiconductor switches [45, 46]. They focused on NMs and associated-technologies performance limits and advantages, when exploited for digital and analog applications, focusing on the main figures of merit needed to meet industry requirements. They discussed using 2D NMs as an enabling factor for flexible electronics and provided perspectives on future developments. Quantum engineering entails electronic-devices' atom-by-atom design and fabrication. The innovative technology, which unifies materials science and device engineering, was fostered by the recent progress in the fabrication of vertical and lateral heterostructures of 2D materials, and assessment of the technology potential *via* computational nanotechnology. However, how close are people to the possibility of the practical realization of next-generation atomically thin transistors? They analyzed the outlook and challenges of quantum-engineered transistors *via* 2D-materials heterostructures, *vs*. Si-technology benchmark and foreseeable evolution in terms of potential performance and manufacturability [47]. Transistors based on lateral heterostructures emerged as the most promising option from a performance viewpoint, even if heterostructure formation and control were in the initial technology development stage.

GR proved to be attractive for thin film (TF) transistors (TFTs) because of its remarkable electronic, optical, mechanical, and thermal properties [48]. Even its major drawback (zero bandgap) resulted in something positive: a resurgence of interest in 2D semiconductors (e.g., TMDs, buckled NMs with sizeable bandgaps). With the discovery of BN-*h* as an ideal dielectric, NMs are in place to advance integrated flexible nanoelectronics, which uniquely take advantage of the unmatched portfolio of 2D-crystals properties, beyond the capability of conventional TFs for ubiquitous flexible systems. The last decades witnessed an extraordinary rise in research progress on ultra-thin 2D NMs in CMP, materials science and chemistry after GR exfoliation from graphite (2004). The unique class of NMs showed many unprecedented properties and is explored for numerous promising applications. Zhang reviewed the state of the art in the development of ultra-thin 2D NMs and highlighted their unique advantages [49]. He discussed the synthetic methods and promising applications of ultra-thin 2D NMs, and personal insights on the challenges in the research area. On the basis of the current achievement on ultra-thin 2D NMs, he gave personal perspectives on potential future research directions. Organic thermoelectrics composed of crystalline low-dimensional (LD) molecular metals present a promising alternative to existing material concepts, especially considering ongoing research to take full control over the amount of charge-transfer and band filling [50]. The peculiar phenomena observed in the material class, e.g., violation of Wiedemann-Franz (WF) law, the anomalous temperature dependence of the electrical conductivity or phonon drag effects in the thermopower, will furthermore open possibilities in low-temperature (LT) thermoelectrics. Manipulating a quantum state *via* electrostatic gating was important for many nanoelectronic model systems. However, controlling the electron spins or, specifically, magnetism of a system by electric-field (EF) tuning proved challenging. Atomically thin magnetic semiconductors attracted attention because of emerging physical phenomena. However, issues are unresolved to show gate-controllable magnetism in 2D materials. Han group showed, *via* electrostatic gating, a strong EF effect in devices based on few-layered ferromagnetic (FM) semiconducting $Cr_2Ge_2Te_6$ [51]. At different gate doping, micro-area Kerr measurements in devices showed bipolar tunable magnetization loops below Curie temperature, which was attributed to the moment rebalance in the spin-polarized band structure. Their findings of EF-controlled magnetism in VDW magnets showed possible applications in magnetic memory storage, sensors, and spintronics.

4.5   A ROADMAP TO IMPLEMENTING MOFS IN ELECTRONIC DEVICES: CHALLENGES

Yaghi group presented hydrothermal synthesis as a route to accessing crystalline, open MOFs, showing extended channel systems and composed of uncommon metal co-ordination. They described the synthesis, structure, and properties of the extended net in crystalline $Cu(4,4'-bpy)_{1.5} \cdot NO_3(H_2O)_{1.25}$ (4,4'-bpy = 4,4'-bipyridine) [52]. They showed that hydrothermal synthesis is a viable route to accessing zeolite-like materials in crystalline form, and presenting components that are not generally observed otherwise. They aimed the application of the synthetic method to the production of similar MOFs, presenting larger channels. They reported the synthesis, structural characterization and gas sorption isotherm measurements for Zn(BDC) (BDC = 1,4-benzenedicarboxylate) microporous MOF of crystalline $Zn(BDC) \cdot (DMF)(H_2O)$ (DMF = *N,N'*-dimethylformamide) [53]. They showed the feasibility of producing microporous MOFs with pores accessible in a zeolite-like fashion, which was expected to pave the way to understanding the chemical and physical properties of the pores, and their ultimate use for molecular transformations and confinement.

Open MOFs are promising materials for applications in catalysis, separation, gas storage and molecular recognition. Compared to conventionally used microporous inorganic materials, e.g., zeolites, open MOFs present the potential for more flexible rational design, *via* control of the architecture and functionalization of the pores. The inability of open MOFs to support permanent porosity and avoid collapsing in the absence of guest molecules, e.g., solvents, hindered further progress in the field. They reported the synthesis of an MOF, which remains crystalline, as evidenced by X-ray single-crystal analyses, and stable when fully desolvated and heated up to 300°C [54]. The synthesis is achieved by borrowing ideas from metal carboxylate cluster chemistry, where an organic dicarboxylate O_2-C-C_6H_4-C-O_2 (BDC) linker is used in a reaction that gives supertetrahedron clusters when capped with monocarboxylates. The rigid and divergent character of the added linker allows the articulation of the clusters into a 3D MOF, resulting in a structure with higher apparent surface area and pore volume than most porous crystalline zeolites. The simple and potentially universal design strategy was pursued in phases and composites synthesis, and for gas-storage applications.

The MOFs represent a class of hybrid organic-inorganic supramolecular materials, comprised of ordered nets formed from organic electron donor linkers and metal cations. They exhibit extremely high surface areas, and tunable pore size and functionality, and act as hosts for a variety of guest molecules. Since their discovery, MOFs enjoyed extensive exploration with applications ranging from gas storage to drug delivery to sensing. Allendorf group reviewed MOF advances, focusing on applications (e.g., gas separation, catalysis, drug delivery, optical, and electronic applications, sensing) [55]. They summarized work on methods for MOF synthesis and computational modeling. The MOFs and related material classes attracted attention for applications (e.g., gas storage, separations, catalysis). In contrast, research focused on potential uses in electronic devices is in its infancy. Several sensing concepts, in which MOFs' tailorable chemistry is used to enhance sensitivity or provide chemical specificity, were shown but in only a few cases, MOFs are an integral part of an actual device. The synthesis of a few electrically conducting MOFs and their known structural flexibility suggest that MOF-based electronic devices, exploiting the properties, could be constructed. However, fabrication methods are required to take advantage of the unique MOFs properties, and extend their use to the realms of electronic circuitry. They described the basic functional elements needed to fabricate electronic devices, summarized relevant MOF research, and reviewed work in which MOFs served as active components in electronic devices [56]. They proposed a high-level roadmap for device-related MOF research, which objective is to stimulate thinking in MOF community, concerning the development MOFs for applications (e.g., sensing, photonics, and microelectronics). The TFs and continuous films of porous MOFs can be conformally deposited on various substrates, *via* a vapor-phase synthesis approach that departs from conventional solution-based routes [57].

4.6 HYBRID POLYMER/MOF FILMS FOR COLORIMETRIC H$_2$O SENSING

Integrating MOFs in microelectronics presents disruptive potential, owing to the unique properties of the microporous crystalline materials. Suitable film deposition methods are crucial to leverage MOFs in the field. Conventional solvent-based procedures, typically adapted from powder preparation routes, are incompatible with nanofabrication owing to corrosion and contamination risks. Ameloot group showed an MOF-chemical vapor deposition (CVD) that enabled high-quality films of prototypical MOF zeolitic imidazole framework (ZIF)-8, with a uniform and controlled thickness even on high-aspect-ratio features [58]. They showed how MOF-CVD enabled previously inaccessible routes (e.g., lift-off patterning, depositing MOF films on fragile features). The compatibility of MOF-CVD with existing infrastructure, in research and production facilities, will greatly facilitate MOF integration in microelectronics. The MOF-CVD was the first vapor-phase deposition method for any type of microporous crystalline net solid, and marked a milestone in processing such materials. The MOFs are typically highlighted for their potential application in gas storage, separations, and catalysis. In contrast, the unique prospects the porous and crystalline materials offer for application in electronic devices, although actively developed, are underexposed. They highlighted the research aimed at MOFs implementation as an integral part of solid-state microelectronics [59]. Manufacturing the devices will critically depend on MOFs compatibility with existing fabrication protocols, and predominant standards. It is important to focus in parallel on a fundamental understanding of MOFs' distinguishing properties, and eliminating fabrication-related obstacles for integration. The latter implies a shift from the microcrystalline powder synthesis in chemistry laboratories to film deposition and processing in a cleanroom environment. They discussed the fundamental and applied aspects of the two-pronged approach. They proposed critical directions for future research in an updated high-level roadmap, to stimulate the next steps towards MOF-based microelectronics in the community.

The MOFs are highly ordered, functionally tunable supramolecular materials with the potential to improve dye-sensitized solar cells (DSSCs) in MOF-DSSCs (MSSCs). Photocurrent could be generated in Grätzel-type DSSC devices when MOFs are used as the sensitizer. However, the specific role(s) of the incorporated MOFs and potential influence of residual MO precursors (MOPs) on device performance are unclear. Allendorf group described the assembly and characterization of a simplified DSSC platform, in which isolated MOF crystals were used as the sensitizer in planar device architecture [60]. They selected a pillared porphyrin framework (PPF) as MOF sensitizer, avoiding contamination from light-absorbing MOPs. Photovoltaic (PV) and electrochemical characterization under simulated 1-sun and wavelength (WL)-selective illumination revealed photocurrent generation, which is ascribable to PPF MOF. Refinement of highly versatile MOF structure and chemistry holds promise for dramatic improvements in emerging PV technologies. The efficiency of DSSCs is strongly influenced by dye molecule orientation and interactions with the substrate. Understanding the factors controlling the surface orientation of sensitizing organic molecules aids in the improvement of traditional DSSCs and devices that integrate molecular linkers at interfaces. They described a general approach to understand relative dye-substrate orientation and provided analytical expressions predicting orientation [61]. They considered the effects of substrate, solvent, and protonation state (pH) on dye molecule orientation. In the absence of solvent, their model predicted that most carboxylic acid (–COOH)-functionalized molecules preferred to lie flat (parallel) on the surface, because of VDW interactions, as opposed to a tilted orientation with respect to the surface, which is favored by -COOH covalent bonding to the substrate. When solvation effects were considered, however, the molecules were predicted to orient perpendicular to the surface. They extended the approach to help understand and guide the orientation of MOF TF growth on various metal-oxide substrates. They developed a two-part analytical model based on density functional theory (DFT) and *ab initio* molecular dynamics (MD, AIMD) simulations results, which predicts the binding energy of a molecule by chemical and dispersion forces on TiO_2-rutile and -anatase surfaces, and quantifies the dye solvation energy for acetonitrile and H_2O. The model is in good agreement with DFT, and enables fast prediction of dye molecule and MOF-linker's binding preference based on the adsorbing-molecule size, surface identity, and solvent environment. They established the threshold molecular size, governing dye molecule orientation for every condition.

Ni_3(2,3,6,7,10,11-hexaiminotriphenylene)$_2$ [Ni_3(HITP)$_2$] is a π-stacked layered MOF with extended π-conjugation, which is analogous to GR. Experiments indicate that the material is semiconducting but theoretical studies predict the bulk material to be metallic. Given that experiment was performed on specimens containing complex nanocrystalline microstructures, and the tendency for internal interfaces to introduce transport barriers, they applied DFT to investigate the influence of internal interface defects on Ni_3(HITP)$_2$ electronic structure [62]. The results showed that interface defects introduced a transport barrier, breaking the π-conjugation and/or decreasing the dispersion of the electronic bands near Fermi level. They showed that the presence of defects could open a small gap in 15–200 meV, which was consistent with the experimentally inferred hopping barrier. Because of their extraordinary surface areas and tailorable porosity, MOFs present the potential to be excellent sensors of gas-phase analytes. The MOFs with open metal sites are particularly attractive for detecting Lewis basic atmospheric analytes, e.g., H_2O. They showed that MOF $Cu_3(BTC)_2$ (Cu-BTC, BTC = 1,3,5-benzenetricarboxylate, HKUST-1) TFs could be used to determine quantitatively the relative humidity (RH) of air *via* a colorimetric approach [63]. The HKUST-1 TFs are spin-coated on to rigid or flexible substrates, and are shown to determine quantitatively RH = 0.1–5% by visual observation or a straightforward optical reflectivity measurement. At high RH > 10%, they used a polymer/MOF bilayer to slow H_2O transport to MOF TF, enabling quantitative determination of RH *via* time as the distinguishing metric. They combined the sensor with an inexpensive LED light source and Si photodiode detector, to show a quantitative RH detector for low RH environments.

4.7 MOF BEATEN TRACK: UNUSUAL STRUCTURES AND UNCOMMON APPLICATIONS

The MOFs received attention in catalysis because of their special structures and a number of promising applications. One of MOFs' intriguing uses is their utilization as precursors for synthesis of metal oxide NMs. Based on the strategy, Gholami group applied Cu-MOF to prepare a series of noble metal NPs (Ag/Au)-decorated CuO (M/CuO, M = Ag, Au) as catalyst [64]. As-prepared nanocomposites (NCs) were characterized by a number of analytical techniques, and their catalytic performances appraised *via* the catalytic reduction of *p*-nitrophenol to *p*-aminophenol as a reliable model reaction. Their results suggested that synergistic effects between Ag/Au and CuO, and high surface area of CuO as appropriate support for NPs elevated catalytic performance of M/CuO materials. One of MOFs usages is their application in manufacturing of metal oxides and metal oxide NCs. They used ZIF-8 as precursor to prepare an Ag/C/ZnO catalyst [65]. The as-prepared samples were characterized *via* x-ray diffraction (XRD), Fourier transform infrared (FTIR), scanning (SEM) and transmission (TEM) electron microscopy analysis. They confirmed improved catalytic performance of as-prepared Ag/C/ZnO by reduction of organic pollutants [e.g., *p*-nitrophenol, methylene blue (MB), rhodamine-B] in the presence of $NaBH_4$ in aqueous media. They proposed Ag-ZnO synergistic effect as the main reason for the remarkably enhanced catalytic activity of Ag/C/ZnO. The MOFs proved themselves as strong contenders in the world of porous materials, standing alongside established compounds classes (e.g., zeolites, activated-C). Following extensive investigation into the porosity of the materials and their gas uptake properties, MOF community branches away from the heavily researched areas and ventures into unexplored avenues. Ranging from novel synthetic routes to MOFs' post-synthetic functionalization, host-guest properties to sensing abilities, Easun group took a sidestep away from increasingly *traditional* approaches in the field, and detailed the most curious qualities of MOFs family [66].

The MOFs are well known for their high surface area and were extensively studied for their gas adsorption properties. Doonan group pioneered an emerging area in MOF science termed *biomimetic mineralization* [67]. The strategy described a facile, *one-pot* approach to encapsulating biomacromolecules within MOF crystals. They showed that the method is highly versatile, presenting proteins, metalloenzymes, and deoxyribonucleic acid (DNA) encapsulation in different MOF materials. They found that MOF architecture protected the biomacromolecule from harsh external environments, allowing it to be exposed to conditions that would lead to its decomposition and deactivation. They employed ZIF-8 to encase biomacromolecules, simply mixing its components, 2-methylimidazole and $Zn(acetate)_2$, in aqueous solution along with the selected biomacromolecule. Small-angle x-ray scattering (SAXS) and FTIR experiments established that the protective capacity of the crystalline coating was attributed to the biomacromolecule's tight encapsulation and ZIF Zn^{2+}–protein-surface interactions. They showed the potential of the novel biocomposites for application to industrial, heterogeneous biocatalysis. Martínez Pérez-Cejuela group informed the use of bio-MOFs, e.g., functionalized *cyclo* dextrines (CDs), as efficient sorbents for solid-phase extraction (SPE) of hydrophilic vitamins (Vits): eight Vits-B (B_1, B_2, B_3, B_5, B_6, B_7, B_8, B_{12}) and C [68]. Bio-MOFs are different from MOFs in that they use biomolecules as ligands. They form inclusion compounds with many molecules. They provided the following conclusions. 1) A bio-MOF for selective recognition of Vit-B_6 was synthesized. 2) An economic, simple, and reliable SPE protocol for Vit-B complex inclusion was developed. 3) The proposed strategy represents a promising alternative for the extraction of hydrophilic Vits.

Much emphasis is placed on reliably obtaining functional materials, especially nanostructured or molecular ones. Most solutions, although usually offering technologically satisfying results, approach the problem from a *top-down* perspective, which is the case for most doped systems from heterogeneous catalysts to conductive polymers. With a background in solution thermodynamics and supramolecular chemistry, Savastano group research goal was to show that a *bottom-up* approach, fine enough to provide control of the prepared materials at the molecular/atomic scale, was a viable option [69]. They discussed heterogeneous catalysts and the need to transition from supported NPs to single-ion catalysts as the ultimate way to maximize per-atom efficiency, and to do so producing ordered systems, presenting chemically well-defined active sites (ASs). Their results in collaboration with Supramolecular Chemistry Group of the Institut Universitari de Ciència Molecular at the Universitat de València, showing the point for organic reactions and electrochemical processes, were presented. They analyzed anion co-ordination chemistry, showing how they transposed detailed understanding of supramolecular interactions in solution to the preparation of novel supramolecular materials. They showed the possibility of obtaining iodine-based SOFs, MOF-like soft-matter, and how to influence their properties in a crystal engineering fashion. Technological relevance (and beauty) of the chemistry was illustrated as they organized increasingly complex nets of polyiodides, which they envisaged as a novel class of molecular conductors and semiconductors. Application of a pillared-layer Zn-triazolate MOF was informed in the dispersive miniaturized SPE of personal care products from wastewater samples [70].

4.8   CONDUCTIVE TWO-DIMENSIONAL (2D) MOFS AS MULTIFUNCTIONAL MATERIALS

Large π-conjugated conductive 2D MOFs, *via* the *bottom-up* approach, emerged as interesting 2D NMs. Du group showed the design and bottom-up fabrication of a noble-metal-free Ni phthalocyanine-based 2D (NiPc-MOF), for highly efficient H_2O oxidation catalysis [71]. The NiPc-MOF was easily grown on various substrates. The TF deposited on F-doped tin oxide (FTO) showed high catalytic O_2 evolution reaction (OER) activity with a low onset potential (<1.48 V, overpotential < 0.25 V), high mass activity

$(883.3 \ A \cdot g^{-1})$ and excellent catalytic durability. Catalyst of 2D MOF TF represented the best OER catalytic activity among molecular catalyst-based NMs. Their work represented the first synthesis of an NiPc-MOF NM and application of π-conjugated 2D MOF for H_2O oxidation. Livingston group informed polymer nanofilms (NFs) with enhanced microporosity by interfacial polymerization [72]. They suggested an evolutionary algorithm (EA), targeted upon the discovery of optimal structures and properties for molecular materials. Conductive 2D MOFs emerged as an only class of multifunctional NMs, because of compositional and structural diversity accessible *via* bottom-up self-assembly. Mirica group reviewed progress in the development of conductive 2D MOFs, with emphasis on synthetic modularity, ND integration strategies, and multifunctional properties [73]. They discussed applications spanning sensing, catalysis, electronics, energy conversion, and storage. They addressed challenges and future outlook, in the context of molecular engineering and practical development of conductive 2D MOFs.

4.9 CALCULATION RESULTS

Typical 2D NMs were measured with layer-layer separation between layers: graphite, black P (P-black), TMD MoS_2, BN-*h*, and an MOF. Separations result lesser than 3.5Å $\ll$ 5 nm. Constancy results in the separation for graphite, P-black, BN-*h*, and MOF. Notwithstanding, the separation results the greatest for MoS_2 owing to its layer-layer arrangement, in which a plane of Mo atoms is inserted in others of S^{2-}. The three strata form an ML of MoS_2. The 2D-NMs periodic table of the elements (PTE) was calculated. It shows [B-N, P, S, Mo] with oxidation states (OSs) 6, 3, 5, –3, etc. Again, a steadiness results in separations for B-N atoms (BN-*h*, graphite) in period 2. Some quantitative structure-property relationships (QSPRs) were gotten: A linear fit of separations *vs.* atomic number Z of 2D NMs was carried out. It results that the separation increases with Z:

$$d = (3.32 \pm 0.04) + (0.00454 \pm 0.00183)Z \tag{1}$$

$$N = 6 \quad r = 0.778 \quad s = 0.057 \quad F = 6.2$$

Notwithstanding, layer-layer separation for S ($Z = 16$, MoS_2) diverges from fit owing to three-strata ML of MoS_2. Slope decays four-fold if only group-V elements (N, P) are included in the trend line:

$$d = 3.32 + 0.001250Z \tag{2}$$

Three-parameter power model of layer-layer separations *vs. Z* was performed. It betters fit:

$$d = 0.1693 + 3.02Z^{0.025} \tag{3}$$

$$N = 6 \quad r = 0.819 \quad s = 0.047$$

Again, layer-layer separation for S (MoS_2) deviates from model owing to three-strata ML of MoS_2.

Two-parameter power model of layer-layer separations *vs. Z* was performed. It betters fit:

$$d = (3.19 \pm 1.02)Z^{(0.0244 \ 0.0085)} \tag{4}$$

$$N = 6 \quad r = 0.820 \quad s = 0.015 \quad F = 8.2$$

One more time, layer-layer separation for S drifts from model owing to three-strata ML of MoS_2.

Possible modes of charge transport (CT) were proposed in 2D MOFs: hopping CT, through-space CT, and through-bond CT. Core organic ligand building blocks were obtained for the construction of 2D MOFs. Triphenylene-based organic ligands were achieved with cross-linking heteroatom (O, S, NH). Table 4.2 lists progress in the field of conductive 2D MOFs with applications (e.g., chemical sensing, FETs, energy conversion, energy storage, catalysis). General methods of integrating conductive 2D MOFs into NDs were found. Applications of traditional and conductive 2D MOFs were gained.

4.10 DISCUSSION

Nanoscience is a general term used to specify the level of knowledge, deep to 1 nm = 10^{-9} m. Together with the technology enabling the application of nanoscience, in a variety of fields [e.g., science, technique, health, society, culture (*nanotechnology*)], they define the Nanoera (new times of unprecedented progress in the human life). Research for materials with applications in technology or healthcare is an important task. The efforts are justified by the need of special properties (e.g., light, and strong, biodegradable, and low-polluting materials and processes).

Interestingly, MOFs are becoming promising materials because of the existence of both inorganic (semiconducting, etc.) and organic (π–π stacking, etc.) properties.

TABLE 4.2 List of Electrically Conducting 2D MOFs with Name and Application

MOF	Electrical Conductivity	Application	Performance
Cu_3HHTP_2[a]	2×10^{-3} S·cm^{-1}	Chemiresistors	Alcohols, aliphatics, amines, aromatics, ketones, ethers, study at 200 ppm
Cu_3HITP_2[b]	2×10^{-1} S·cm^{-1}		NH_3 – LOD[f] 0.5 ppm; range of study 0.5–10 ppm
Ni_3HITP_2	2S·cm^{-1}		NH_3 – LOD 0.5 ppm; range of study 1–100 ppm
Cu_3HITP_2	2×10^{-1} S·cm^{-1}	Chemiresistors	Range of study 5–80 ppm, H_2S-LOD 0.23 ppm, NO-LOD 1.4 ppm
Cu_3HHTP_2	0.02 S·cm^{-1}	Chemiresistors	H_2S-LOD 0.52 ppm, NO-LOD 0.16 ppm
Ni_3HHTP_2	2.8 MΩ·cm^{-2}	Chemiresistors	Range of study 5–80 ppm, array
Ni_3HITP_2	5.6 MΩ·cm^{-2}		– H_2S-LOD 35 ppm,
Fe_3HHP_2/graphite	3.2×10^{-2} S·cm^{-1}	Chemiresistors	array-NH_3-LOD 19 ppm,
Co_3HHT_2/graphite	9.8×10^{-1} S·cm^{-1}		array-NO-LOD 17 ppm
Ni_3HHTP_2/graphite	3.8×10^{-2} S·cm^{-1}		
Cu_3HHTP_2/graphite	2.8×10^{-1} S·cm^{-1}		
Cu_3HHTP_2	7.6×10^{-3} S·cm^{-1}	Chemiresistors	Range of study 2.5–80 ppm, array-H_2S LOD 40 ppm, array-NO LOD 40 ppm
Ni_3HHTP_2	1.0×10^{-2} S·cm^{-1}		
Cu_3HHTP_2	–	Electrochemical ion sensing	11.1 μA·h^{-1}, K$^+$ – LOD
Ni_3HHTP_2	–		6.31×10^{-7} M,
Co_3HHTP_2	–		NO_3^- – LOD 5.01×10^{-7} M
CoBHT[c]	–	Hydrogen evolution	Overpotential, solution 0.5 M H_2SO_4, pH = 1.3, –0.19 V
NiBHT	–		–0.33 V
FeBHT	–		–0.47 V
THTNi 2DSP	–	Hydrogen evolution	Overpotential, solution 0.5 M H_2SO_4, –0.33 V
[Co$_3$(BHT)$_2$]$^{3-}$	–	Hydrogen evolution	Overpotential, solution H_2SO_4, pH = 1.3, –0.34 V
[Co$_3$(THT)$_2$]$^{3-}$	–		–0.53 V
Cu-BHT	280 S·cm^{-1}	Hydrogen evolution	Overpotential, solution H_2SO_4, –0.45 V or –0.76 V

TABLE 4.2 *(Continued)*

MOF	Electrical Conductivity	Application	Performance
NiAT	3×10^{-6} S·cm^{-1}	Hydrogen evolution	Overpotential, solution H_2SO_4, pH = 1.3, -0.37 V
NiPc-MOF[d]	0.2 S·cm^{-1}	Water oxidation	Low onset, <1.48 V, overpotential <0.25 V
Ni$_3$HITP$_2$	40 S·cm^{-1}	Oxygen reduction	Onset 0.82 V, overpotential 0.18 V
Ni$_3$HITP$_2$	40 S·cm^{-1}	Supercapacitors	Capacitance, 111 F·g^{-1}, 18 mF·cm^{-2}
Cu$_3$HHTP$_2$	3×10^{-1} S·cm^{-1}	Supercapacitors	capacitance, 120 F·g^{-1}, 22 mF·cm^{-2}
Ni$_3$HAB$_2$	—	Supercapacitors	Capacitance, 420 F·g^{-1}, Ni = 1.6 F·cm^{-2} Ni = 760 F·cm^{-3}
Cu$_3$HAB$_2$	—		Capacitance, 215 F·g^{-1}; Cu = 0.86 F·cm^{-2}; Cu −
Co$_3$HTTP$_2$[e]	3.4×10^{-9} S·cm^{-1}	Electrochemical capture of ethylene	Ethylene captured, 126.8 mmol·g^{-1}
Ni$_3$HTTP$_2$	3.6×10^{-4} S·cm^{-1}		218.2 mmol·g^{-1}
Cu$_3$HTTP$_2$	2.4×10^{-8} S·cm^{-1}		100.2 mmol·g^{-1}
Ni$_3$HITP$_2$	40 S·cm^{-1}	Field-effect transistors	48.6 cm^2·V^{-1}·s^{-1}, p-type
M-HIB	—	Field-effect transistors	Slight variation in conductance with variation of a back-gate voltage
Cu-BHT	1×10^3 S·cm^{-1}	Field-effect transistors	116 cm^2·V^{-1}·s^{-1}

[a] HHTP: 2,3,6,7,10,11-hexahydroxytriphenylene.
[b] HITP: hexaiminotriphenylene.
[c] BHT: benzenehexathiolate.
[d] Pc: phthalocyanine.
[e] HTTP: hexathiotriphenylene.
[f] LOD: limit of detection.

Source: Ko, Mendecki, and Mirica [73].

4.11 FINAL REMARKS

From the present results and discussion, the following final remarks can be drawn:

1. Metal-organic frameworks are new materials with an infinity of applications.
2. Imagination is the only limit that people have.
3. The most important and most difficult question for any science is classification.
4. Nanoscience vital law results: Contrast models with reality.
5. A biometal-organic framework for selective recognition of vitamin-B_6 was synthesized. An economic, simple, and reliable SPE protocol for vitamin-B complex inclusion was developed. The proposed strategy represents a promising alternative for the extraction of hydrophilic vitamins.
6. It was investigated how best to integrate experiments, computation, and theory. Science is seen from outside as experimental but another theoretical science exists, and classification is a part of theoretical science.
7. Interestingly, MO frameworks are becoming promising materials because of the existence of both inorganic (semiconducting, etc.) and organic (π–π stacking, etc.) properties.

ACKNOWLEDGMENTS

The authors thank the support from Generalitat Valenciana (Project No. PROMETEO/2016/094) and Universidad Católica de Valencia *San Vicente Mártir* (Project No. UCV.PRO.17–18.AIV.03).

KEYWORDS

- **bio-MOF**
- **charge transport**
- **functional material**

- **host-guest chemistry**
- **monolayer**
- **periodic table**
- **post-synthetic modification**
- **quantitative structure-property relationship**
- **thin film**
- **thin-layered nanostructure**
- **transition metal dichalcogenide**

REFERENCES

1. Feynman, R. P., (1960). There is plenty of room at the bottom. *Caltech Eng. Sci.*, *23*, 22–36.
2. Anderson, P. W., (1972). More is different: Broken symmetry and the nature of the hierarchical structure of science. *Science*, *177*, 393–396.
3. Novoselov, K. S., Geim, A. K., Morozov, S. V., Jiang, D., Zhang, Y., Dubonos, S. V., Grigorieva, I. V., & Firsov, A. A., (2004). Electric field effect in atomically thin carbon films. *Science*, *306*, 666–669.
4. Britnell, L., Ribeiro, R. M., Eckmann, A., Jalil, R., Belle, B. D., Mishchenko, A., et al., (2013). Strong light-matter interactions in heterostructures of atomically thin films. *Science*, *340*, 1311–1314.
5. Coronado, E., Martí-Gastaldo, C., Navarro-Moratalla, E., Ribera, A., Blundell, S. J., & Baker, P. J., (2010). Coexistence of superconductivity and magnetism by chemical design. *Nat. Chem.*, *2*, 1031–1036.
6. Pinilla-Cienfuegos, E., Kumar, S., Mañas-Valero, S., Canet-Ferrer, J., Catala, L., Mallah, T., Forment-Aliaga, A., & Coronado, E., (2015). Imaging the magnetic reversal of isolated and organized molecular-based nanoparticles using magnetic force microscopy. *Part. Part. Syst. Charact.*, *32*, 693–700.
7. Pinilla-Cienfuegos, E., Mañas-Valero, S., Forment-Aliaga, A., & Coronado, E., (2016). Switching the magnetic vortex core in a single nanoparticle. *ACS Nano*, *10*, 1764–1770.
8. Gott, J. A., Beanland, R., Fonseka, H. A., Zhang, Y., Liu, H., & Sanchez, A. M., (2017). Stable step facets in III-V semiconducting nanowires. *Microsc. Anal.*, *33*, 12–18.
9. Torrens, F., (2003). Effect of elliptical deformation on molecular polarizabilities of model carbon nanotubes from atomic increments. *J. Nanosci. Nanotech.*, *3*, 313–318.
10. Torrens, F., (2004). Effect of size and deformation on polarizabilities of carbon nanotubes from atomic increments. *Future Generation Comput. Syst.*, *20*, 763–772.
11. Torrens, F., (2004). Effect of type, size, and deformation on polarizability of carbon nanotubes from atomic increments. *Nanotechnology*, *15*, S259–S264.

12. Torrens, F., (2006). Corrigendum: Effect of type, size, and deformation on polarizability of carbon nanotubes from atomic increments. *Nanotechnology, 17*, 1541–1541.

13. Torrens, F., (2004). Periodic table of carbon nanotubes based on the chiral vector. *Internet Electron. J. Mol. Des., 3*, 514–527.

14. Torrens, F., (2005). Periodic properties of carbon nanotubes based on the chiral vector. *Internet Electron. J. Mol. Des., 4*, 59–81.

15. Torrens, F., (2005). Calculations on *cyclo* pyranoses as co-solvents of single-wall carbon nanotubes. *Mol. Simul., 31*, 107–114.

16. Torrens, F., (2005). Calculations on solvents and co-solvents of single-wall carbon nanotubes: *Cyclo* pyranoses. *J. Mol. Struct. (Theochem.), 757*, 183–191.

17. Torrens, F., (2005). Calculations on solvents and co-solvents of single-wall carbon nanotubes: *Cyclo* pyranoses. *Nanotechnology, 16*, S181–S189.

18. Torrens, F., (2005). Some calculations on single-wall carbon nanotubes. *Probl. Nonlin. Anal. Eng. Syst., 11*(2), 1–16.

19. Torrens, F., (2006). Calculations of organic-solvent dispersions of single-wall carbon nanotubes. *Int. J. Quantum Chem., 106*, 712–718.

20. Torrens, F., & Castellano, G., (2005). Cluster origin of the solubility of single-wall carbon nanotubes. *Comput. Lett., 1*, 331–336.

21. Torrens, F., & Castellano, G., (2007). Cluster nature of the solvation features of single-wall carbon nanotubes. *Curr. Res. Nanotech., 1*, 1–29.

22. Torrens, F., & Castellano, G., (2007). Effect of packing on the cluster nature of C nanotubes: An information entropy analysis. *Microelectron. J., 38*, 1109–1122.

23. Torrens, F., & Castellano, G., (2007). Cluster origin of the transfer phenomena of single-wall carbon nanotubes. *J. Comput. Theor. Nanosci., 4*, 588–603.

24. Torrens, F., & Castellano, G., (2007). Asymptotic analysis of coagulation-fragmentation equations of carbon nanotube clusters. *Nanoscale Res. Lett., 2*, 337–349.

25. Torrens, F., & Castellano, G., (2008). Properties of fullerite and other symmetric carbon forms: Similarity laws. *Symmetry Cult. Cult. Sci., 19*, 341–370.

26. Torrens, F., & Castellano, G., (2010). Fullerite crystal thermodynamic characteristics and the law of corresponding states. *J. Nanosci. Nanotechn., 10*, 1208–1222.

27. Torrens, F., & Castellano, G., (2010). Cluster nature of the solvent features of single-wall carbon nanohorns. *Int. J. Quantum Chem., 110*, 563–570.

28. Torrens, F., & Castellano, G., (2011). (Co-)solvent selection for single-wall carbon nanotubes: Best solvents, acids, super acids and guest-host inclusion complexes. *Nanoscale, 3*, 2494–2510.

29. Torrens, F., & Castellano, G., (2012). Bundlet model for single-wall carbon nanotubes, nanocones and nanohorns. *Int. J. Chemoinf. Chem. Eng., 2*(1), 48–98.

30. Torrens, F., & Castellano, G., (2013). Solvent features of cluster single-wall C, BC_2N and BN nanotubes, cones, and horns. *Microelectron. Eng., 108*, 127–133.

31. Torrens, F., & Castellano, G., (2013). Corrigendum to: Solvent features of cluster single-wall C, BC_2N and BN nanotubes, cones, and horns. *Microelectron. Eng., 112*, 168–168.

32. Torrens, F., & Castellano, G., (2013). Bundlet model of single-wall carbon, BC_2N and BN nanotubes, cones, and horns in organic solvents. *J. Nanomater. Mol. Nanotech., 2*, Article 1000107, 1–9.

33. Torrens, F., & Castellano, G., (2013). C-nanostructures cluster models in organic solvents: Fullerenes, tubes, buds and graphenes. *J. Chem. Chem. Eng., 7*, 1026–1035.

34. Torrens, F., & Castellano, G., (2014). Cluster solvation models of carbon nanostructures: Extension to fullerenes tubes and buds. *J. Mol. Model.*, *20*, Article 2263, 1–9.

35. Torrens, F., & Castellano, G., (2014). Cluster model expanded to C-nanostructures: Fullerenes, tubes, graphenes and their buds. *Austin J. Nanomed. Nanotech.*, *2*(2), Article 7, 1–7.

36. Torrens, F., & Castellano, G., (2013). Elementary polarizability of Sc/fullerene/graphene aggregates and di/graphene-cation interactions. *J. Nanomater. Mol. Nanotech.*, *S1*, Article 001, 1–8.

37. Torrens, F., & Castellano, G., (2018). Conductive layered metal-organic frameworks: A chemistry problem. *Int. J. Phys. Study Res.*, *1*(2), 42–42.

38. Torrens, Z. F., (2019). From layered materials to bidimensional metal-organic frameworks. *Nereis*, *11*, 63–78.

39. Torrens, F., & Castellano, G., (2019). Conductive two-dimensional nanomaterials: Metal-organic frameworks. *J. Appl. Phys. Nanotech.*, *2*(2), 52–52.

40. Torrens, F., Castellano, G. (2018). Conductive layered metal-organic frameworks: A chemistry problem. *Int. J. Phys. Study Res.,* *1*(2), 42.

41. Wang, Q. H., Kalantar-Zadeh, K., Kis, A., Coleman, J. N., & Strano, M. S., (2012). Electronics and optoelectronics of two-dimensional transition metal dichalcogenides. *Nature Nanotech.*, *7*, 699–712.

42. Fukagawa, H., Shimizu, T., Hanashima, H., Osada, Y., Suzuki, M., & Fujikake, H., (2012). Highly efficient and stable red phosphorescent organic light-emitting diodes using platinum complexes. *Adv. Mater.*, *24*, 5099–5103.

43. Butler, S. Z., Hollen, S. M., Cao, L., Cui, Y., Gupta, J. A., Gutiérrez, H. R., et al., (2013). Progress, challenges, and opportunities in two-dimensional materials beyond graphene. *ACS Nano*, *7*, 2898–2926.

44. Meric, I., Dean, C. R., Petrone, N., Wang, L., Hone, J., Kim, P., & Shepard, K. L., (2013). Graphene field-effect transistors based on boron-nitride dielectrics. *Proc. IEEE*, *101*, 1609–1619.

45. Fiori, G., Bonaccorso, F., Iannaccone, G., Palacios, T., Neumaier, D., Seabaugh, A., Banerjee, S. K., & Colombo, L., (2014). Electronics based on two-dimensional materials. *Nat. Nanotech.*, *9*, 768–779.

46. Fiori, G., Bonaccorso, F., Iannaccone, G., Palacios, T., Neumaier, D., Seabaugh, A., Banerjee, S. K., & Colombo, L., (2014). Electronics based on two-dimensional materials. *Nat. Nanotech.*, *9*, 1063–1063.

47. Iannaccone, G., Bonaccorso, F., Colombo, L., & Fiori, G., (2018). Quantum engineering of transistors based on 2D materials heterostructures. *Nat. Nanotechnol.*, *13*, 183–191.

48. Akinwande, D., Petrone, N., & Hone, J., (2014). Two-dimensional flexible nanoelectronics. *Nat. Commun.*, *5*, 5678–5678.

49. Zhang, H., (2015). Ultrathin two-dimensional nanomaterials. *ACS Nano*, *9*, 9451–9469.

50. Huewe, F., Steeger, A., Kostova, K., Burroughs, L., Bauer, I., Strohriegl, P., Dimitrov, V., Woodward, S., & Pflaum, J., (2017). Low-cost and sustainable organic thermoelectrics based on low-dimensional molecular metals. *Adv. Mater.*, *29*, Article 1605682, 1–7.

51. Wang, Z., Zhang, T., Ding, M., Dong, B., Li, Y., Chen, M., et al., (2018). Electric-field control of magnetism in a few-layered van der Waals ferromagnetic semiconductor. *Nat. Nanotechn.*, *13*, 554–559.

52. Yaghi, O. M., & Li, H., (1995). Hydrothermal synthesis of a metal-organic framework containing large rectangular channels. *J. Am. Chem. Soc., 117*, 10401–10402.

53. Li, H., Eddaoudi, M., Groy, T. L., & Yaghi, O. M., (1998). Establishing microporosity in open metal-organic frameworks: Gas sorption isotherms for Zn(BDC) (BDC = 1,4-benzenedicarboxylate). *J. Am. Chem. Soc., 120*, 8571–8572.

54. Li, H., Eddaoudi, M., O'Keeffe, M., & Yaghi, O. M., (1999). Design and synthesis of an exceptionally stable and highly porous metal-organic framework. *Nature (London), 402*, 276–279.

55. Meek, S. T., Greathouse, J. A., & Allendorf, M. D., (2011). Metal-organic frameworks: A rapidly growing class of versatile nanoporous materials. *Adv. Mater., 23*, 249–267.

56. Allendorf, M. D., Schwartzberg, A., Stavila, V., & Talin, A. A., (2011). A roadmap to implementing metal-organic frameworks in electronic devices: Challenges and critical directions. *Chem. Eur. J., 17*, 11372–11388.

57. Allendorf, M. D., & Stavila, V., (2016). Nanoporous films: From conventional to conformal. *Nat. Mater., 15*, 255–257.

58. Stassen, I., Styles, M., Grenci, G., van Gorp, H., Vanderlinden, W., De Feyter, S., Falcaro, P., De Vos, D., Vereecken, P., & Ameloot, R., (2016). Chemical vapor deposition of zeolitic imidazolate framework thin films. *Nat. Mater., 15*, 304–310.

59. Stassen, I., Burtch, N., Talin, A., Falcaro, P., Allendorf, M., & Ameloot, R., (2017). An updated roadmap for the integration of metal-organic frameworks with electronic devices and chemical sensors. *Chem. Soc. Rev., 46*, 3185–3241.

60. Spoerke, E. D., Small, L. J., Foster, M. E., Wheeler, J., Ullman, A. M., Stavila, V., Rodriguez, M., & Allendorf, M. D., (2017). MOF-sensitized solar cells enabled by a pillared porphyrin framework. *J. Phys. Chem. C., 121*, 4816–4824.

61. Domenico, J., Foster, M. E., Spoerke, E. D., Allendorf, M. D., & Sohlberg, K., (2018). Effect of solvent and substrate on the surface binding mode of carboxylate-functionalized aromatic molecules. *J. Phys. Chem. C., 122*, 10846–10856.

62. Foster, M. E., Sohlberg, K., Allendorf, M. D., & Talin, A. A., (2018). Unraveling the semiconducting/metallic discrepancy in $Ni_3(HITP)_2$. *J. Phys. Chem. Lett., 9*, 481–486.

63. Ullman, A. M., Jones, C. G., Doty, F. P., Stavila, V., Talin, A. A., & Allendorf, M. D., (2018). Hybrid polymer/metal-organic framework films for colorimetric water sensing over a wide concentration range. *ACS Appl. Mater. Interfaces, 10*, 24201–24208.

64. Akbarzadeh, E., Falamarzi, M., & Gholami, M. R., (2017). Synthesis of M/CuO (M = Ag, Au) from Cu based metal organic frameworks for efficient catalytic reduction of *p*-nitrophenol. *Mater. Chem. Phys., 198*, 374–379.

65. Falamarzi, M., Akbarzadeh, E., & Gholami, M. R., (2019). Zeolitic imidazolate framework-derived Ag/C/ZnO for rapid reduction of organic pollutant. *J. Iran. Chem. Soc., 16*, 1105–1111.

66. Tansell, A. J., Jones, C. L., & Easun, T. L., (2017). MOF the beaten track: Unusual structures and uncommon applications of metal-organic frameworks. *Chem. Cent. J., 11*, Article 100, 1–16.

67. Doonan, C. J. (2019). Personal communication.

68. Martínez Pérez-Cejuela, H., Mon, M., Ferrando-Soria, J., Pardo, E., Simó-Alfonso, E. F., & Herrero-Martínez, J. M. (2019). Personal communication.

69. Savastano, M. (2019). Personal communication.
70. González-Hernández, P., Lago, A. B., Pasán, J., Ruiz-Pérez, C., Ayala, J. H., Afonso, A. M., & Pino, V., (2019). Application of a pillared-layer Zn-triazolate metal-organic framework in the dispersive miniaturized solid-phase extraction of personal care products from wastewater samples. *Molecules*, *24*, Article 690, –1–17.
71. Jia, H., Yao, Y., Zhao, J., Gao, Y., Luob, Z., & Du, P., (2018). A novel two-dimensional nickel phthalocyanine-based metal-organic framework for highly efficient water oxidation catalysis. *J. Mater. Chem. A.*, *6*, 1188–1195.
72. Jimenez-Solomon, M. F., Song, Q., Jelfs, K. E., Munoz-Ibanez, M., & Livingston, A. G., (2016). Polymer nanofilms with enhanced microporosity by interfacial polymerization. *Nat. Mater.*, *15*, 760–767.
73. Ko, M., Mendecki, L., & Mirica, K. A., (2018). Conductive two-dimensional metal-organic frameworks as multifunctional materials. *Chem. Commun.*, *54*, 7873–7891.

CHAPTER 5

Periodic Table of Elements: Heavy, Rare, Critical, and Superelements

FRANCISCO TORRENS[1] and GLORIA CASTELLANO[2]

[1]*Institute for Molecular Science, University of Valencia, PO Box 22085, E-46071, Valencia, Spain, E-mail: torrens@uv.es*

[2]*Department of Experimental Sciences and Mathematics, Faculty of Veterinary and Experimental Sciences, Valencia Catholic University Saint Vincent Martyr, Guillem de Castro-94, E-46001 Valencia, Spain*

ABSTRACT

The contingent and socially constructed character of toxicity are investigated to reveal mechanisms via which the diverse protagonists, under the pressure of strong political, economic, and academic interests, interact to visualize the toxicity of a substance or to make it invisible, via the active production of ignorance about its effects on health and the environment. The instances illustrate that poisonous danger organization results by powerful inequities among creators and injured parties, which promote the growth of types of sluggish and more often than not imperceptible aggression against communally deprived groupings. A number of accounts of the periodic table give an arresting prompt that persons require to use Refuse-Reduce-Reuse-Repurpose-Recycle (five R's) at a basic height. It is simple to take for granted, other than up till now no option exists to show, that plain-body atoms result in complex creatures shaped by the arrangement of a number of even now smaller fractions, etc. The cyclic habit exposed by Mendeleev among the features and mass corroborates this forewarning. Researchers find out 26 transuranics in Mendeleev's periodic table. Notwithstanding, the boundary of the periodic table is even now distant. Physicists suppose that superheavy elements must be with so much long existence that they could be found out in the Cosmos.

5.1   INTRODUCTION

Setting the scene: the periodic table of the elements (PTE), heavy, rare, critical, superelements, toxic metals, (in)visible toxics, the emergence of the heavy elements, conductive two-dimensional (2D) metal-organic (MO) framework, rare-earth elements fingerprinting anthropogenic activities and driving chemical processes in archaeology, PTE's endangered elements, conservation of critical elements, the way to superelements, and enigmas in the periodic law (PL).

Through examples, the contingent and socially constructed character of toxicity was investigated to reveal some of the mechanisms through which the diverse protagonists, under the pressure of strong political, economic, and academic interests, interact to visualize the toxicity of a substance or, on the contrary, to make it invisible, through the active production of ignorance about its effects on health and the environment. The examples also show that toxic risk management is marked by strong imbalances between producers and victims, which foster the development of forms of slow and usually invisible violence against socially disadvantaged groups. A quantity of descriptions of the periodic system offers an outstanding aide-memoire that folks call for relating Refuse-Reduce-Reuse-Repurpose-Recycle (five R's) at a primary stage. It is trouble-free to suppose, however, there is so far no likelihood to reveal, that minimal-body atoms are difficult creatures modeled by the organization of a quantity of in spite of everything lesser divisions, and so on. The periodic custom uncovered by Mendeleev between the characteristics and weight verifies the presentiment. Scientists learned 26 transuranics in Mendeleev's periodic table. Nevertheless, the bound of the periodic table is in spite of everything remote. Physicists take for granted that superheavy elements should exist with so much drawn out life that they could be learned in the World. The formal structure of the periodic system of the chemical elements (PTE) was informed [1].

In earlier publications, it was reported PTE [2–4], quantum simulators [5–13], science, and ethics of developing sustainability via nanosystems and devices [14], *green nanotechnology* as an approach towards environment safety [15], molecular devices, machines as hybrid organic-inorganic structures [16] PTE, quantum biting its tail and sustainable chemistry [17]. Back to PTE? In the present report, it is discussed PTE, heavy, rare, critical, superelements, toxic metals, (in)visible toxics, the emergence of the heavy elements, conductive 2D MO framework, rare-earth PTE fingerprinting anthropogenic activities and driving chemical processes in archaeology, PTE's endangered elements, conservation of critical elements, the way to superelements and enigmas in the PL. There is a general interest in toxic metals, (in)visible toxics, PTE's endangered elements, conservation of critical elements and the way to superelements. The aim of this work is to initiate a debate by suggesting a number of questions (Q), which can arise when addressing subjects of (in)visible toxics, the way to superelements and enigmas in the PL. It was provided, when possible, answers (A), hypotheses (H) and facts (F).

5.2 TOXIC METALS

A narrow relationship exists between biosphere evolution and PTE: really, all organisms, since their first evolutionary stages, made use of chemical properties of many metal ions for the development of their essential biochemical functions [18]. As a consequence, the metallic elements, even in low doses, are essential for the development of their vital functions. Their insufficient contribution leads to organisms-developmental anomalies; however, when their concentration is higher to optimum, they exert toxic effects on organisms, limiting their development. Cadmium is found in PTE in the same column as Hg and Zn, but its chemical properties are closer to Zn, which similarity affects Cd distribution and its toxic properties. Arsenic and P, found in the same PTE column, present a similar chemical behavior; however, oxidation state (OS) reduction 5 to 3 results easier in Arsenic.

5.3 (IN)VISIBLE TOXICS

Universitat de València organized the exposition *(In)visible Toxics*, raising questions (*cf.* Figure 5.1).

Q1. Which are the most dangerous toxins?
Q2. Which is their degree of toxicity?
Q3. What risks do they pose to people or to the environment?
Q4. What measures should be adopted to avoid damage and minimize future risks?
Q5. *Via* what mechanisms do actors interact to visualize substance toxicity or make it invisible?

The following conclusions (Cs) were provided:

C1. Toxicity's contingent/socially constructed character revealed mechanisms *via* which protagonists interact to visualize substance toxicity/make it invisible.

C2. Examples show that strong producers-victims imbalances mark toxic risk management.

TÒXICS (IN)VISIBLES

Del 2 d'abril al 28 de juny de 2019

La rectora de la Universitat de València es complau a convidar-vos a la inauguració de l'exposició «Tòxics (in)visibles» que tindrà lloc al Palau de Cerveró el 2 d'abril, a les 19:00 hores.

www.uv.es/cultura/exposicions

VNIVERSITAT ID VALÈNCIA
PALAU de CERVERÓ

Universitat d'Alacant
Universidad de Alicante
UNIVERSITAS
Miguel Hernández

FIGURE 5.1   Exposition *(In)visible toxics*.

5.4   THE EMERGENCE OF THE HEAVY ELEMENTS

Heavy PTE Nb, Mo, Ru-Xe, Cs-Nd, Sm-Rn, and Fr-U (OSs 3, 4, 2, 1, 5, 6, etc.) were formed in merging neutron stars (NSs, *cf.* Figure 5.2).

FIGURE 5.2   In emergence of some heavy PTEs, most were formed merging NSs: Pt/Au factory in sky.

5.5 CONDUCTIVE TWO-DIMENSIONAL (2D) METAL-ORGANIC (MO) FRAMEWORKS

Some 2D materials PTE [B-N, P, S, Mo] (*cf.* Figure 5.3) shows OSs 6, 3, 5, –3, etc. [19].

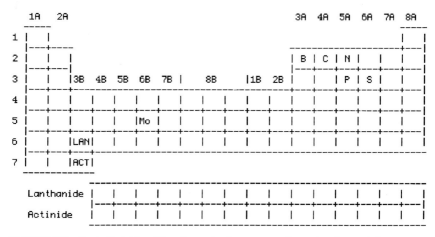

FIGURE 5.3 Periodic table of the elements of some two-dimensional materials.

5.6 RARE-EARTH ELEMENTS FINGERPRINTING ANTHROPOGENESIS/DRIVING PROCESSES

Rare-earth PTE fingerprints anthropogenic activities and drives chemical processes in archaeology: [Sc] (OS 3, *cf.* Figure 5.4, *black*), [Y] (OS 3, *red*), light lanthanoids [La-Gd] (OS 3, *green*) and heavy lanthanoids [Tb-Lu] (OS 3, *blue*) [20].

5.7 THE PTE'S ENDANGERED ELEMENTS: CONSERVATION OF CRITICAL ELEMENTS

The periodic table's endangered elements follow: limited availability and future risk to supply [Li, B, Mg, P, Sc, V-Mn, Co-Cu, Se, Sr-Mo, Pd, Cd, Sn, Sb, W, Au-Bi, Nd] (OS 2, 3, 5, 4, etc., *cf.* Figure 5.5, *black*); rising threat from increased use [Ru, Rh, Ta, Os-Pt, U] (OS 4, 3, 5, 6, etc., *orange*); serious threat in the next 100 years [He, Zn-As, Ag, In, Te, Hf] (OS 3, 4, etc., *red*).

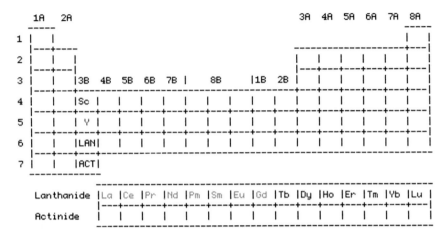

FIGURE 5.4   PTE fingerprints anthropogenesis: Sc (*black*); Y (*red*); light/heavy lanthanoids (*green/blue*).

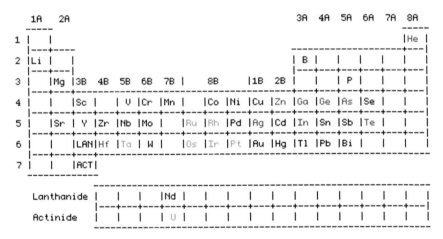

FIGURE 5.5   PTE's endangered elements: Limited availability (*black*); rising/serious threats (*orange/red*).

The PTE of conservation of critical elements shows remaining years until known reserves are depleted (based on current rate of use): 5–20 years [Zn, Se, Sn, Sb, Au, Tl, Pb] (OS 2, 3, 4 and 1, *cf.* Figure 5.6, *red*), 20–50 years [Mn, Ga-As, Sr, Ru, Rh, Ag-In, Hf, W, Os-Pt, Bi, U] (OS 3, 4, 2, 6 and 1, *orange*), 50–100 years [P, S, Co-Cu, Zr-Mo, Pd, Te, Ta, Re, Hg, Np] (OS 2, 5, 4, 6 and 7, *blue*) and 100–500 years [He-B, Mg, Al, Sc, V, Cr, Y, I, Ce-Lu] (OS 3, 2, –1, 0, 1 and 5, *black*) [21].

```
     1A   2A                                                3A  4A  5A  6A  7A  8A
     ----                                                   -----------------------
  1 | |   |                                                                    |He |
    |---+----|                                              ----------------------+---|
  2 |Li |Be |                                               | B |   |   |   |   |   |
    |---+---|                                               |---+---+---+---+---+---|
  3 |   |Mg |3B  4B  5B  6B  7B |     8B      |1B  2B |Al |   | P | S |   |   |
    |---+---+---+---+---+---+---+---+---+---+---+---+---+---+---+---+---+---+---|
  4 |   |   |Sc |   | V |Cr |Mn |   |Co |Ni |Cu |Zn |Ga |Ge |As |Se |   |   |
    |---+---+---+---+---+---+---+---+---+---+---+---+---+---+---+---+---+---+---|
  5 |   |Sr | Y |Zr |Nb |Mo |   |Ru |Rh |Pd |Ag |Cd |In |Sn |Sb |Te | I |   |
    |---+---+---+---+---+---+---+---+---+---+---+---+---+---+---+---+---+---+---|
  6 |   |   |LAN|Hf |Ta | W |Re |Os |Ir |Pt |Au |Hg |Tl |Pb |Bi |   |   |   |
    |---+---+---+---+---------------------------------------------------------|
  7 |   |   |ACT|
    ---------------

 Lanthanide |   |Ce |Pr |Nd |Pm |Sm |Eu |Gd |Tb |Dy |Ho |Er |Tm |Yb |Lu |
            |---+---+---+---+---+---+---+---+---+---+---+---+---+---+---|
 Actinide   |   |   |   | U |Np |   |   |   |   |   |   |   |   |   |   |
```

FIGURE 5.6 Remaining years of reserves: 5–20 (*red*); 20–50 (*orange*); 50–100 (*blue*); 100–500 (*black*).

A new version of PTE shows the elements under threat (*cf.* Figure 5.7) [22]. Element (e.g., Ag, He, Sr, OS 1, 0, 2) are the most under thread (*red, and many play an important part in smartphones*). The PTE shows (*black*) elements that could be more ethically sourced (*conflict minerals*) because wars are fought and lives lost over their ownership.

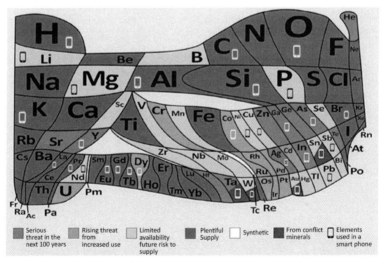

FIGURE 5.7 A new version of the periodic table shows the elements under threat.
Image credit: European Chemical Society.
Reprinted from https://theanalyticalscientist.com/fields-applications/red-alert

5.8   THE PERIODIC TABLE OF C-X BONDS IN BIOLOGY

The periodic table of C-X bonds in biology follow: [H, C-F, P-Cl, Fe-Ni, As-Br, I] (OS – 1, 2, 3, 4, etc., *cf.* Figure 5.8).

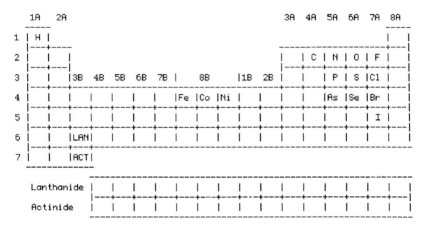

FIGURE 5.8   The periodic table of C-X bonds in biology.

5.9   METALLOFULLERENES WITH ELEMENTS OF THE PERIODIC TABLE

The PTE of metallofullerenes shows elements [Os-Pt] (OS 4 and 2, *cf.* Figure 5.9) [23].

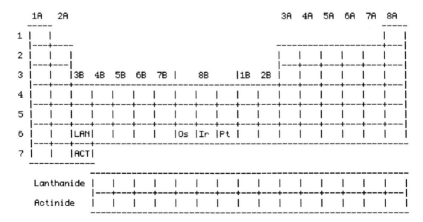

FIGURE 5.9   Metallofullerenes with elements of the periodic table of the elements.

5.10   THE PERIODIC TABLE OF POLYOXOMETALLATES (POMS)

The PTE of polyoxometallates (POMs) shows elements [B, Al-P, Ga-As] (OS 3, 4 and 5, *cf.* Figure 5.10) [24].

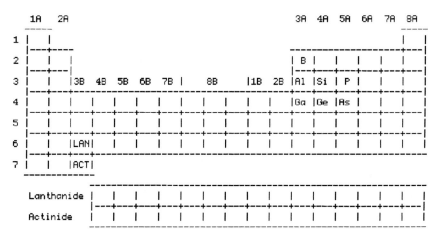

FIGURE 5.10   Polyoxometallates with elements of the periodic table of the elements.

5.11   IN THE WAY TO SUPERELEMENTS: ENIGMAS IN THE PERIODIC LAW (PL)

Fliórov and Ilínov proposed questions, A, H, and F and on enigmas in the PL [25].

Q1.   How during the last years did scientists discover 26 transuranics in PTE?

Q2.   What does the surrounding world consist of?

Q3.   Why around people do so many different substances find that are in gaseous/liquid/solid states?

Q4.   Is it possible or not to turn some substances into others or obtain new ones, unknown before?

Q5.   The chemical properties of elements, are they fortuitous or change according to a certain law?

A5.   (Mendeleev). PL: The properties of the elements change periodically *vs.* atomic weight.

Q6.   What are the causes of periodicity with which the properties of the elements change?

H1. (Mendeleev). It is easy to assume, but there is yet no possibility to demonstrate, that simple-bodies atoms are complicated creatures formed by the composition of some still more small parts... The periodic dependence revealed by me between the properties and weight confirms such premonition.

F1. (Rutherford). Atom represents a complicated system in which center a positively charged nucleus is, around which negatively charged electrons rotate in different orbits.

Q7. Where is the limit of the periodic table?

Q8. (Mendeleev). How many elements does the periodic system contain?

A8. Calculations show that when $Z = 170–180$, the nucleus should capture the inner electrons.

Q9. (Mendeleev). Where does it pass its limit?

A9. Such limit is not determined by electronic layer instability but by that of the proper nucleus.

5.12   FINAL REMARKS

From the presents results and discussion, the following final remarks can be drawn:

1. Through examples, the contingent and socially constructed character of toxicity was investigated to reveal some of the mechanisms through which the diverse protagonists, under the pressure of strong political, economic, and academic interests, interact to visualize the toxicity of a substance or, on the contrary, to make it invisible, through the active production of ignorance about its effects on health and the environment.

2. The examples also show that toxic risk management is marked by strong imbalances between producers and victims, which foster the development of forms of slow and usually invisible violence against socially disadvantaged groups.

3. Some versions of the periodic table provide a striking reminder that people need to apply Refuse-Reduce-Reuse-Repurpose-Recycle (five Rs) at a fundamental level.

4. During the last years, scientists discovered 26 transuranics in Mendeleev's periodic table. However, the limit of the periodic

system is still far. Physicists assume that superheavy elements should exist with so much lengthy life that they could be discovered in the Universe.

ACKNOWLEDGMENTS

The authors thank the support from Generalitat Valenciana (Project No. PROMETEO/2016/094) and Universidad Católica de Valencia *San Vicente Mártir* (Project No. 2019-217-001UCV).

KEYWORDS

- **anthropogenic activity**
- **archaeology**
- **conductive two-dimensional metal-organic framework**
- **critical-element conservation**
- **endangered element**
- **heavy-element emergence**
- **invisible toxic**
- **periodic law enigma**
- **rare earth**
- **superelements way**
- **toxic metal**
- **visible toxic**

REFERENCES

1. Leal, W., & Restrepo, G., (2019). Formal structure of periodic system of elements. *Proc. R. Soc. A., 475*, Article 20180581, 1–20.
2. Torrens, F., & Castellano, G., (2015). Reflections on the nature of the periodic table of the elements: Implications in chemical education. In: Seijas, J. A., Vázquez, T. M. P., & Lin, S. K., (eds.), *Synthetic Organic Chemistry* (Vol. 18, pp. 1–15). MDPI: Basel, Switzerland.

3. Torrens, F., & Castellano, G., (2018). Nanoscience: From a two-dimensional to a three-dimensional periodic table of the elements. In: Haghi, A. K., Thomas, S., Palit, S., & Main, P., (eds.), *Methodologies and Applications for Analytical and Physical Chemistry* (pp. 3–26). Apple Academic-CRC: Waretown, NJ.

4. Torrens, F., & Castellano, G. (2019). Periodic table. In: Putz, M. V., (ed.), New Frontiers in Nanochemistry: Concepts, Theories, and Trends. Apple Academic-CRC, Waretown, NJ, Vol. 1, 403–425.

5. Torrens, F., & Castellano, G., (2015). Ideas in the history of nano/miniaturization and (Quantum) simulators: Feynman, education and research reorientation in translational science. In: Seijas, J. A., Vázquez, T. M. P., & Lin, S. K., (eds.), *Synthetic Organic Chemistry* (Vol. 19, pp. 1–16). MDPI: Basel, Switzerland.

6. Torrens, F., & Castellano, G., (2015). Reflections on the cultural history of nano-miniaturization and quantum simulators (computers). In: Laguarda, M. N., Masot, P. R., & Brun, S. E., (eds.), *Sensors and Molecular Recognition* (Vol. 9, pp. 1–7). Universidad Politécnica de Valencia: València, Spain.

7. Torrens, F., & Castellano, G., (2016). Nanominiaturization and quantum computing. In: Costero, N. A. M., Parra, Á. M., Gaviña, C. P., & Gil, G. S., (eds.), *Sensors and Molecular Recognition* (Vol. 10, Chapter 31, pp. 1–5). Universitat de València: València, Spain.

8. Torrens, F., & Castellano, G., (2018). Nanominiaturization, classical/quantum computers/simulators, superconductivity and universe. In: Haghi, A. K., Thomas, S., Palit, S., & Main, P., (eds.), *Methodologies and Applications for Analytical and Physical Chemistry* (pp. 27–44). Apple Academic-CRC: Waretown, NJ.

9. Torrens, F., & Castellano, G., (2018). Superconductors, superconductivity, BCS theory, and entangled photons for quantum computing. In: Haghi, A. K., Aguilar, C. N., Thomas, S., & Praveen, K. M., (eds.), *Physical Chemistry for Engineering and Applied Sciences: Theoretical and Methodological Implication* (pp. 379–387). Apple Academic-CRC: Waretown, NJ.

10. Torrens, F., & Castellano, G., (2018). EPR paradox, quantum decoherence, qubits, goals and opportunities in quantum simulation. In: Haghi, A. K., (ed.), *Theoretical Models and Experimental Approaches in Physical Chemistry: Research Methodology and Practical Methods* (Vol. 5, pp. 317–334). Apple Academic-CRC: Waretown, NJ.

11. Torrens, F., & Castellano, G. (2019). Nanomaterials, molecular ion magnets, ultra-strong and spin-orbit couplings in quantum materials. In: Vakhrushev, A. V., Haghi, R., & de Julián-Ortiz, J. V., (eds.), *Physical Chemistry for Chemists and Chemical Engineers: Multidisciplinary Research Perspectives* (pp. 181–190). Apple Academic-CRC: Waretown, NJ.

12. Torrens, F., & Castellano, G. (2019). Nanodevices and organization of single-ion magnets and spin qubits. In: Balköse, D., Ribeiro, A. C. F., Haghi, A. K., Ameta, S. C., & Chakraborty, T., (eds.), *Chemical Science and Engineering Technology: Perspectives on Interdisciplinary Research* (pp. 67–74). Apple Academic-CRC: Waretown, NJ.

13. Torrens, F., & Castellano, G. (2019). Superconductivity and quantum computing via magnetic molecules. In: Haghi, A. K., (ed.), *New Insights in Chemical Engineering and Computational Chemistry* (pp. 201–209). Apple Academic-CRC: Waretown, NJ.

14. Torrens, F., & Castellano, G. (2019). Developing sustainability via nanosystems and devices: Science-ethics. In: Balköse, D., Ribeiro, A. C. F., Haghi, A. K., Ameta, S. C., &

Chakraborty, T., (eds.), *Chemical Science and Engineering Technology: Perspectives on Interdisciplinary Research* (pp. 75–84). Apple Academic-CRC: Waretown, NJ.

15. Torrens, F., & Castellano, G. (2020). Green nanotechnology: An approach towards environment safety. In: Vakhrushev, A. V., Ameta, S. C., Susanto, H., & Haghi, A. K., (eds.), *Advances in Nanotechnology and the Environmental Sciences: Applications, Innovations, and Visions for the Future* (pp. 85–91). Apple Academic-CRC: Waretown, NJ.

16. Torrens, F., & Castellano, G. (2020). Molecular devices/machines: Hybrid organic-inorganic structures. In: Pourhashemi, A., Deka, S. C., & Haghi, A. K., (eds.), *Research Methods and Applications in Chemical and Biological Engineering* (pp. 13–24). Apple Academic-CRC: Waretown, NJ.

17. Torrens, F., & Castellano, G. (2020). The periodic table, quantum biting its tail, and sustainable chemistry. In: Torrens, F., Haghi, A. K., & Chakraborty, T., (eds.), *Chemical Nanoscience and Nanotechnology: New Materials and Modern Techniques* (pp. 25–32). Apple Academic-CRC: Waretown. NJ.

18. Spiro, T. G., & Stigliani, W. M., (2003). *Chemistry of the Environment*. Prentice-Hall: Upper Saddle River, NJ.

19. Ko, M., Mendecki, L., & Mirica, K. A., (2018). Conductive two-dimensional metal-organic frameworks as multifunctional materials. *Chem. Commun., 54*, 7873–7891.

20. Gallello, G. (2019). Personal communication.

21. Supanchaiyamat, N., & Hunt, A. J., (2019). Conservation of critical elements of the periodic table. *Chem. Sus. Chem., 12*, 1–8.

22. The Analytical Scientist, (2019). Red alert! A new version of the periodic table shows the elements under threat. *Anal. Scientist, 73*, 9.

23. Martín, N. (2019). Personal communication.

24. Poblet, J. M. (2019). Personal communication.

25. Fliórov, G. N., & Ilínov, A. S., (1985). *En el Camino Hacia Los Superelementosp.* Mir: Moscow.

CHAPTER 6

Periodic Table, Chemical Bond Nature, and Nonclassical Compounds

FRANCISCO TORRENS[1] and GLORIA CASTELLANO[2]

[1]*Institute for Molecular Science, University of Valencia, PO Box 22085, E-46071, Valencia, Spain, E-mail: torrens@uv.es*

[2]*Department of Experimental Sciences and Mathematics, Faculty of Veterinary and Experimental Sciences, Valencia Catholic University Saint Vincent Martyr, Guillem de Castro-94, E-46001 Valencia, Spain*

ABSTRACT

The periodic table of the elements (PTE) is essential to understanding our nature and place in the whole of beings, food, drugs, materials, nanomaterials (NMs), etc. The ideas in the periodic table should be valued by the questions that they generate. A working guide with questions and answers was written to introduce the periodic table in chemical education and provide answers. The periodic table was related to electron configurations. Q1. Without looking at the periodic table but reminding the number of elements in every row (2, 8, 8, 18, 18, 32, 32), which elements have the atomic numbers 9, 10, 11, 17, 19, 35, 37 and 54? *Etcetera*. A need exists to develop the periodic-table research to understand periodic properties and periodic law (PL). The periodic table is empirical: more important than nomenclature are the facts of chemical behavior depending on the outer-core shells and frontier orbitals. It illustrates chemists' ability to summarize information in a two-dimensional (2D) representation. The most striking points of the periodic table continue to be: (1) the elements that occupy the same column present similar chemical properties; (2) the opposed behavior between both extreme groups (*alkaline* elements, *halogens*).

6.1 INTRODUCTION

Setting the scene: the periodic table of the elements (PTE), groups in PTE, working guide and the nature of chemical bonds in the new non-classical compounds and materials.

The PTE is essential to understand our nature and place in the whole of beings, food, drugs, materials, nanomaterials (NMs), etc. The ideas in PTE should be valued by the questions that they generate. A working guide with questions (Q) and answers (A) was written to introduce PTE in chemical education and provide answers. The PTE was related to electron configurations.

Q1. Without looking at PTE but reminding the number of elements in every row (2, 8, 8, 18, 18, 32, 32), which elements have the atomic numbers 9, 10, 11, 17, 19, 35, 37 and 54?

Q2. Can you outline a PTE plan and place from memory the symbols of the first 18 elements and of the remaining alkali metals, halogens, and *inert* elements?

Q3. Consult the melting and boiling points of the alkali metals in a table. Compare these sequences, *vs.* the atomic number, with the corresponding ones to halogens and *inert* elements. Can you suggest an explanation?

Q4. Can you give the names and formulae of four Na compounds?

Etcetera: There is a need to develop PTE research to understand periodic properties and periodic law (PL). The PTE is empirical: more important than nomenclature are the facts of chemical behavior depending on the outer-core shells and frontier orbitals. Nowadays, the most striking points of PTE continue to be: (1) the elements that occupy the same column present similar chemical properties; (2) the opposed behavior between both extreme groups (*alkaline* elements, *halogens*).

In this group, Valero-Molina reported PTE and its relationship with daily life [1]. Earlier publications presented empirical didactics [2–4], resonance in interacting induced-dipole polarizing force fields [5], its implementation in chemical education [6], tool for macromolecular-structure interrogation/retrieval [7], reflections on the cultural history of nanominiaturization and quantum simulators (computers) [8], and the nature of PTE [9–13].

6.2   GROUPS IN THE PERIODIC TABLE OF THE ELEMENTS (PTE)

Table 6.1 shows the groups in the PTE.

TABLE 6.1   Groups in the Periodic Table of the Elements

Group Number	Name
1	alkali metals
2	alkaline earth metals
13	boron
14	carbon
15	nitrogen
16	chalcogens
17	Halogens
18	noble gases
–	transition elements
–	lanthanides
–	actinides

6.3   WORKING GUIDE

Casanova Freixas and del Barrio Estévez raised questions providing answers [14, 15]. We plan activities, neither objectives nor contents to help teachers stimulate inquiry-based learning. We propose thinking and answering individually the following questions. Pending problems can be worked in groups.

Q1.   Do all electrons in an atom occupy the 1s-orbital, which has the least energy?

A1.   Pauli exclusion principle. No, the maximum number of electrons that can occupy the same orbital is two.

Q2.   A gas mass formed by H atoms, would be an element?

A2.   Yes, but such an elementary form based on monoatomic H is difficult to observe *as is* in the Earth. For people, H is a molecular gas of formula H_2.

Q3.   What are two gaseous allotropic forms of O_x molecules?

A3.   The O is O_2 but it also exists ozone O_3. The O_2 and O_3 are allotropic forms of O.

Q4. What is the difference between water and a mixture of H_2 and O_2?

A4. In water, there are only triatomic molecules H-O-H. In H_2/O_2 mixture, molecules H-H and O-O exist.

Q5. What is the deprotonation in water of a weak acid, e.g., HF?

A5. $HF + H_2O \Leftrightarrow F^- + H_3O^+$.

Q6. Without looking at PTE but reminding the number of elements in every row (2, 8, 8, 18, 18, 32, 32), which elements have the atomic numbers 9, 10, 11, 17, 19, 35, 37 and 54 [16]?

Q7. Can you outline a PTE plan and place from memory the symbols of the first 18 elements and of the remaining alkali metals, halogens, and *inert* elements?

Q8. Consult the melting and boiling points of the alkali metals in a table. Compare these sequences, *vs.* the atomic number, with the corresponding ones to halogens and *inert* elements. Can you suggest an explanation?

Q9. Can you give the names and formulae of four Na compounds?

Q10. Can you compare the properties of HCl and NaCl?

Q11. What would be observed if a solution containing a great amount of KI be shaken with a small quantity of $Cl_{2(g)}$?

Q12. Remembering that in the ordinary chemical reactions the atoms are neither created nor destroyed, can you write the adjusted equations for the following reactions: a) hydrogen and oxygen (O_2), to form water; b) sulfuric acid (H_2SO_4) and magnesium, to form hydrogen and magnesium sulfate ($MgSO_4$); c) reduction of cupric oxide (CuO), with hydrogen, to form copper and water; d) hydrogen release by reaction of potassium with water; e) reaction of carbon dioxide (CO_2) with potassium hydroxide, to form potassium carbonate; f) reaction of fluorhydric acid with silica (SiO_2), to form silicon tetrafluoride (SiF_4); g) copper burning in fluorine, to form cupric fluoride (CuF_2); h) reaction of sodium hydroxide with chlorhydric acid, to form kitchen salt; i) reaction of sodium bromide with sulfuric acid, to form hydrogen bromide; j) carbon-dioxide release from sodium hydrogen carbonate with sulfuric acid?

Q13. From their places in PTE, can you predict some properties of At and Fr?

Q14. What are allotropic forms? What differences can be mentioned between the properties and structure of O_2 and O_3?

Q15. Are other elements known, in addition to O and C, which exist in allotropic forms?

Q16. What are the names and chemical formulae of the substances with the following common names: lampblack, quartz, borax, magnesia, limestone, quicklime, slaked lime, gypsum, plaster, barite, emery, ruby, sapphire, and alum?

Q17. Read the list of the names of all the elements. How many are heard about in the ordinary life, apart from the chemistry lessons? Make a list of 10 of the least common elements, and write a sentence establishing some fact on its basis, learned in the ordinary life.

Q18. Make a list of all the generalizations about the variation of the physical and chemical properties with regard to the position in PTE, which could be discovered. During the study of chemistry, remember the generalizations; check these, as well as new ones that could be discovered, in the light of every new studied chemical fact; e.g., What regularity is there in the formulae of the hydrides of group V? Compare with the formulae of the main acids of the elements of group V. Friedel and Maloney raised the following Qs on the subscript in a chemical formula [17].

Q19. How many oxygen atoms are present in a container with 288 g of O_3?

Datum: molar mass of O_3 is 48.0 g.

Q20. There are 1.8×10^5 atoms in a sample of P_4. What is the mass of this sample?

Datum: molar mass of P_4 is 124 g.

Q21. How many atoms of sulfur are in a sample of 963 g of S_6?

Datum: gram atomic weight of S is 32.1 g.

Q22. There are 2.41×10^{24} atoms in a sample of S_8. What is the mass of this sample?

Datum: gram atomic weight of S is 32.1 g.

Cervellini et al. raised the following questions [18].

Q23. How do you classify the elements in accordance with their physical and chemical properties?

Q24. Which are the periodic properties?

Q25. Can metallic character be verified in the practical laboratory work?

6.4 NATURE OF CHEMICAL BONDS IN NON-CLASSICAL COMPOUNDS AND MATERIALS

The computational design of new compounds, which present non-standard structure and unusual properties, is a significant part of modern science. In addition to the classical covalent two-center two-electron (2c–2e) bonds, there are non-classical types of bonds: three-center two-electron (3c–2e), four-center two-electron (4c–2e) and 16-center two-electron (16c–2e) bonds. Figure 6.1 shows PTE [B, C, Al, Ga, In, oxidation states (OSs) 3, –4 and 4] of new compounds that present the non-standard structure and unusual properties of elements.

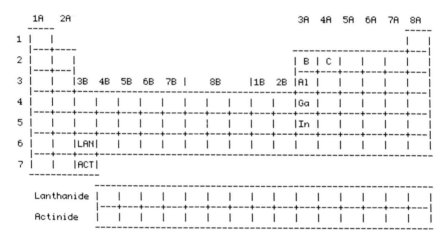

FIGURE 6.1   PTE of new compounds that present the non-standard structure and unusual properties.

6.5 FINAL REMARKS

From the present results and discussion, the following final remarks can be drawn:

1. Mendeleev drew the following conclusions:
   a. The elements, if arranged according to their atomic weights, exhibit an evident *periodicity* of properties.
   b. Elements which are similar as regards their chemical properties have atomic weights which are either of nearly the same

value (e.g., platinum, iridium, osmium) or which increase regularly (e.g., potassium, rubidium, cesium).

c. The arrangement of the elements, or of groups of elements, in the order of their atomic weights, corresponds to their so-called *valencies* as well as, to some extent, to their distinctive chemical properties (as is apparent, among other series, in that of lithium, beryllium, barium, carbon, nitrogen, oxygen, and iron).

d. The elements which are the most widely diffused have *small* atomic weights.

e. The *magnitude* of the atomic weight determines the character of the element, just as the magnitude of the molecule determines the character of a compound.

f. We must expect the discovery of many yet *unknown* elements (e.g., elements analogous to aluminum and silicon, whose atomic weight would be between 65 and 75).

g. The atomic weight of an element may sometimes be amended by knowledge of those of the contiguous elements. Thus, the atomic weight of tellurium must lie between 123 and 126, and cannot be 128.

h. Certain characteristic properties of the elements can be foretold from their atomic weights. [...]

2. Moseley drew the following conclusions:

a. Every element from Al to Au is characterized by an integer N which determines its X-ray spectrum. Every detail in the spectrum of an element can, therefore, be predicted from the spectra of its neighbors.

b. This integer N, the atomic number of the element, is identified with the number of positive units of electricity contained in the atomic nucleus.

c. The atomic numbers for all elements from Al to Au have been tabulated on the assumption that N for Al is 13.

d. The order of the atomic numbers is the same as that of the atomic weights, except where the latter disagrees with the order of the chemical properties.

e. Known elements correspond with all the numbers between 13 and 79 except three. There are here three possible elements still undiscovered.

    f.   The frequency of any line in the X-ray spectrum is approximately proportional to $A/(N-b)^2$, where A and b are constants.

ACKNOWLEDGMENTS

The authors thank the support from Generalitat Valenciana (Project No. PROMETEO/2016/094) and Universidad Católica de Valencia *San Vicente Mártir* (Project No. 2019-217-001UCV).

KEYWORDS

- **atomic number**
- **atomic volume**
- **doubly connected arrangement**
- **element**
- **empirical chemometrics**
- **empirical definition**
- **non-classical material**
- **ordering concept**
- **period**
- **periodic law**
- **periodic system**
- **periodicity**
- **qualitative chemistry**
- **valence**
- **working guide**

REFERENCES

1. Valero-Molina, R., (2013). The periodic system and its relation to everyday life. Part I. *An. Quim, 109*, 301–307.
2. Torrens, F., Sánchez-Pérez, E., & Sánchez-Marín, J., (1989). Empirical teaching of the molecular shape. *Ensenanza de las Ciencias (Extra III Congr.), 1*, 267, 268.

3. Torrens, F., (2000). Phylogeny of anthropoid apes. *Encuentros en la Biologia, 8*(60), 3–5.

4. Torrens, F., (2000). Fractal analysis of the tertiary structure of proteins. *Encuentros en la Biologia, 8*(64), 4–6.

5. Torrens, F., & Castellano, G., (2009). Resonance in interacting induced-dipole polarizing force fields: Application to force-field derivatives. *Algorithms, 2,* 437–447.

6. Torrens, F., & Castellano, G., (2014). A new tool for the study of resonance in chemical teaching. *Avances en Ciencia e Ingenieria, 5*(1), 81–91.

7. Torrens, F., & Castellano, G., (2014). A tool for interrogation of macromolecular structure. *J. Mater. Sci. Eng. B., 4*(2), 55–63.

8. Torrens, F., & Castellano, G., (2015). Reflections on the cultural history of nanominiaturization and quantum simulators (computers). In: *Sensors and Molecular Recognition* (Vol. 9, Chapter 1, pp. 1–7). Universidad Politécnica de Valencia: València, Spain.

9. Torrens, F., & Castellano, G., (2015). Reflections on the nature of the periodic table of the elements: Implications in chemical education. In: Seijas, J. A., Vázquez, T. M. P., & Lin, S. K., (eds.), *Synthetic Organic Chemistry* (Vol. 18, pp. 1–15). MDPI: Basel, Switzerland.

10. Putz, M. V. (2019). *New Frontiers in Nanochemistry: Concepts, Theories, and Trends.* Apple Academic-CRC: Waretown, NJ.

11. Torrens, F., & Castellano, G., (2018). Nanoscience: From a two-dimensional to a three-dimensional periodic table of the elements. In: Haghi, A. K., Thomas, S., Palit, S., & Main, P., (eds.), *Methodologies and Applications for Analytical and Physical Chemistry* (pp. 3–26). Apple Academic-CRC: Waretown, NJ.

12. Torrens, F., & Castellano, G. (2020). The periodic table, quantum biting its tail, and sustainable chemistry. In: Torrens, F., Haghi, A. K., & Chakraborty, T., (eds.), *Chemical Nanoscience and Nanotechnology: New Materials and Modern Techniques* (pp. 25–32). Apple Academic-CRC: Waretown, NJ.

13. Torrens, F., & Castellano, G. Periodic table of elements: Heavy, rare, critical and super elements. In: Haghi, A. K., (ed.), *Theoretical and Empirical Analysis in Physical Chemistry: Framework for Research.* Apple Academic-CRC: Waretown, NJ, in press.

14. Casanova, F. À., & Del Barrio, E. L., (2012). Essential Chemical Dictionary: Definitions, Examples, Exercises, Graphics... + Practical Formulation Summary and Periodic Table of the Elements. Larousse: Barcelona, Spain.

15. Del Barrio, E. L., (2012). Essential Physical Dictionary: Definitions, Examples, Exercises, Graphics... + Practical Summary of Formulae. Larousse: Barcelona, Spain.

16. Pauling, L., (1947). *General Chemistry.* W.H. Freeman & Co.: San Francisco, CA.

17. Friedel, A. W., & Maloney, D. P., (1995). Those baffling subscripts. *J. Chem. Educ., 72,* 899–905.

18. Cervellini, M. I., Chasvin, O. M. N., Muñoz, M. Á., Zambruno, M. A., & Morazzo, G., (2012). Experimental practices in basic chemistry. In: Pinto, C. G. & Martin, S. M., (eds.), *Teaching and Dissemination of Chemistry and Physics* (pp. 455–460). Ibergarceta: Madrid, Spain.

CHAPTER 7

Artemisia integrifolia, A. capillaris Components, and Sesquiterpene Lactones

FRANCISCO TORRENS[1] and GLORIA CASTELLANO[2]

[1] *Institute for Molecular Science, University of Valencia, PO Box 22085, E-46071, Valencia, Spain, E-mail: torrens@uv.es*

[2] *Department of Experimental Sciences and Mathematics, Faculty of Veterinary and Experimental Sciences, Valencia Catholic University Saint Vincent Martyr, Guillem de Castro-94, E-46001 Valencia, Spain*

ABSTRACT

Artemisia integrifolia is a medicinal and edible plant. To investigate its antihyperlipidaemic effect, a crude lipophilic extract and the composing compounds were isolated and fractioned from the petroleum ether extract of *A. integrifolia* aerial parts *via* column chromatography on SiO_2 gel. The antihyperlipidaemia effect was studied in a rat model of acute hyperlipidaemia, which was induced by triton WR-1339. A new compound, integrinol, and nine known compounds [chamazulene, acetylenes (E)-2 and -3, eugenol, and palmitic, oleic, linoleic, linolenic, and 12,13-epoxylinolenic acids] were isolated from *A. integrifolia* crude lipophilic extract. Among sesquiterpene lactones (STLs), the ditriazolyl cumanin derivative results more active and selective than cumanin in the tested breast, cervix, lung, and colon tumor cell lines. The compound is the least toxic *vs.* splenocytes (concentration causing 50% cell death of 524.1 μM) and exhibited the greatest selectivity on tumor cell lines. The compound showed a 50% growth inhibition of 2.3 μM and a selectivity index (SI) of 227.9 on WiDr human colon tumor cell line. The compound can be considered for further studies and is a candidate for the development of antitumor agents.

7.1   INTRODUCTION

Artemisia integrifolia L. (Compositae) is widely distributed in Inner Mongolia, where it is used as a medicinal and edible plant. Pharmacological studies showed that its ethanol extract possesses good superoxide, hydroxyl, and NO free-radical (FR) scavenging activities and can inhibit lipid peroxidation. With regard to its chemical constituents, fatty acids, acetylenes, phenylpropanoids, and terpenoids were isolated from *A. integrifolia* ether extract, which exhibited many pharmacological activities; e.g., eugenol protects the liver from damage by scavenging FRs, increasing the activity of superoxide dismutase (SOD), glutathione peroxidase (GSH-Px) and catalase (CAT), and reducing the synthesis of malondialdehyde (MDA). Linolenic acid can increase eicosapentaenoic (EPA) and docosahexaenoic (DHA)-acids levels in platelets and erythrocytes, and reduce arachidonic-acid production. However, until recently, only few available reports exist on the antihyperlipidaemic activity, identification, and isolation of *A. integrifolia* lipophilic components, all of which may present an impact on the development of novel therapeutic options to treat hyperlipidaemia and related conditions.

It was informed chemical composition and antimicrobial activities of *Artemisia argyi* essential oils (EOs), extracted by simultaneous distillation-extraction (SDE), subcritical extraction and hydrodistillation [1]. It was reported antimicrobial and antioxidant activities of *A. herba-alba* EO cultivated in Tunisian arid zone [2]. The extracts of *A. absinthium* suppress the growth of hepatocellular carcinoma cells *via* induction of apoptosis *via* endoplasmic reticulum (ER) stress and mitochondrial-dependent pathway [3].

Earlier publications classified 31 sesquiterpene lactones (STLs) [4, 5]. It was informed the tentative mechanism of action, resistance of artemisinin (ART) derivatives (ARTDs) [6], reflections, proposed molecular mechanism of bioactivity, resistance [7], chemical, biological screening approaches, phytopharmaceuticals [8], chemical components from *Artemisia austro-yunnanensis*, anti-inflammatory effects, lactones [9], triazole-derived, artesunate, and metabolic pathways for ART [10]. The aim of this work is to review ART, ARTDs, antihyperlipidemic effect, identification, and isolation of the lipophilic components from *A. integrifolia*, simultaneous determination and risk assessment (RA) of pyrrolizidine alkaloids (PAs) in *A. capillaris*, and cytotoxic activity and selectivity of STL derivatives

(STLDs). The purpose of this report is to review recent advances in ART, ARTDs, and STLDs.

7.2   ARTEMISININ AND ITS DERIVATIVES

As a result of the Vietnam War, as the North Vietnamese forces were suffering heavily from malaria, they appealed to their ally, China, for help [11]. Many Chinese herbs were tested and attention was drawn to an ancient herbal remedy based on the plant *Artemisia annua*, which was used by Chinese herbalists for over 1 000 years. Peasants in rural areas made a tea from it (*via* hot, not boiling, H_2O) or chewed fresh plants mixed with brown sugar, as a way of treating fevers. A team led by Professor Tu Youyou obtained ART, the active ingredient, by low-temperature (LT) ether extraction; they found the formula $C_{15}H_{22}O_5$ of the white crystals in 1972 and its structure was determined in 1976. The ART (*cf.* Figure 7.1a) presents relatively short-term potency; because of this, it is not a suitable prophylactic. It also suffers from a lack of H_2O solubility. Semisynthetic ARTDs developed to improve its potency were soon in use as antimalarial drugs (e.g., artesunate, Figure 7.1b, artemether, Figure 7.1c). When *Plasmodium* parasites ingest hemoglobin (Hb) to extract protein for growth, heme groups are freed. Either Haem groups or free Fe^{2+} attack the peroxide group in ART, resulting in the generation of free radicals (FRs) that attack and kill *Plasmodium* deoxyribonucleic acid (DNA). The ART was suggested to inhibit enzyme *Plasmodium falciparum* PfATP6 that is involved in pumping Ca^{2+} into membrane organelles, which may be a means of its action. Concerns exist about the possible emergence of ART-resistant parasites, which is compounded by the widespread availability of fake ART medications, especially in Southeast Asia, and by people not completing their course of treatment. The ART combination therapy (ACT, *via* artemether-lumefantrine, artesunate-mefloquine, artesunate-amodiaquine, artesunate-sulphadoxine/pyrimethamine) was endorsed by the World Health Organization (WHO) as a frontline treatment for severe *P. falciparum* malaria.

The most promising totally synthetic drug to be investigated is OZ277 (Arterolane, *cf.* Figure 7.2). The starting point in its design was the *endoperoxide* bridge that is key to ART activity. The first synthesized compounds were too unstable and not active as a drug, so it was decided to protect the bridge incorporating a bulky adamanty group on one side of the bridge, which made the molecule not just more stable but also better at killing malarial parasites than the best ARTDs. The first compounds of the type were not soluble, so the next step was to incorporate some polar groups to enhance this. The OZ277 is not just more soluble, but also is absorbed well by the body when administered orally, and is effective *vs.* malaria in- *in vitro* and *in vivo* testing. It is now in phase-III human trials and it is believed to act similarly to ART. Antimalarial drugs should not, of course, be seen in isolation. They do not remove the need for other countermeasures (e.g., mosquito nets, covering skin with clothing, use of mosquito repellent).

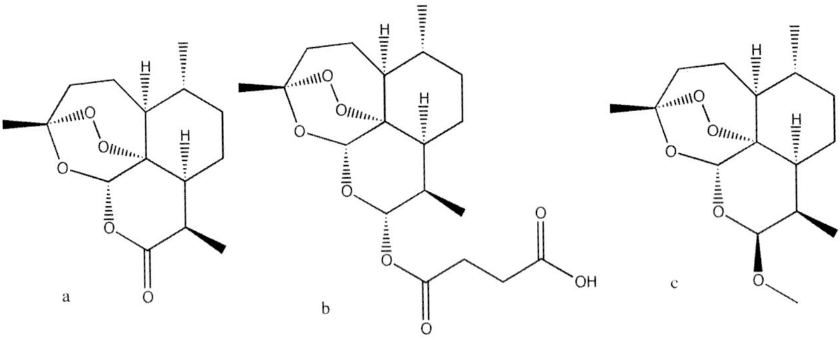

FIGURE 7.1   Molecular structures: a) artemisinin, b) artesunate and c) artemether.

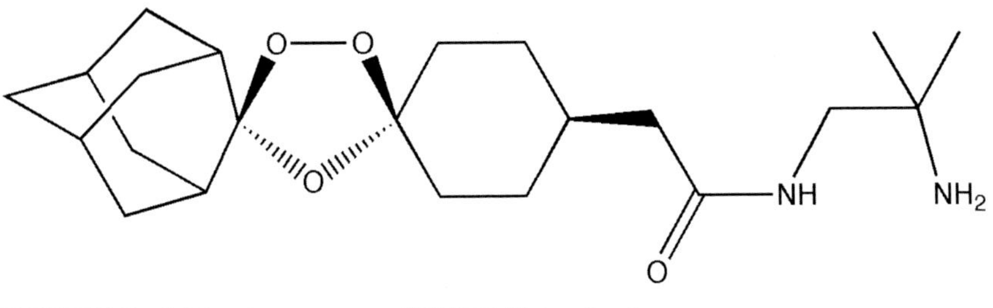

FIGURE 7.2   Molecular structure of OZ277 (Arterolane).

7.3   *A. INTEGRIFOLIA* COMPONENTS ANTIHYPERLIPIDAEMIC/ IDENTIFICATION/ISOLATION

For getting the most useful chemical information, column chromatography and nuclear magnetic resonance (NMR) spectroscopy were used to isolate and identify *Artemisia integrifolia* crude lipophilic components [12]. A new compound, integrinol, was isolated and identified from crude lipophilic components by chromatographic fractionation, together with nine known compounds [chamazulene, acetylenes (E)-2 and -3, eugenol, and palmitic, oleic, linoleic, linolenic, and 12,13-epoxylinolenic acids, *cf.* Figure 7.3].

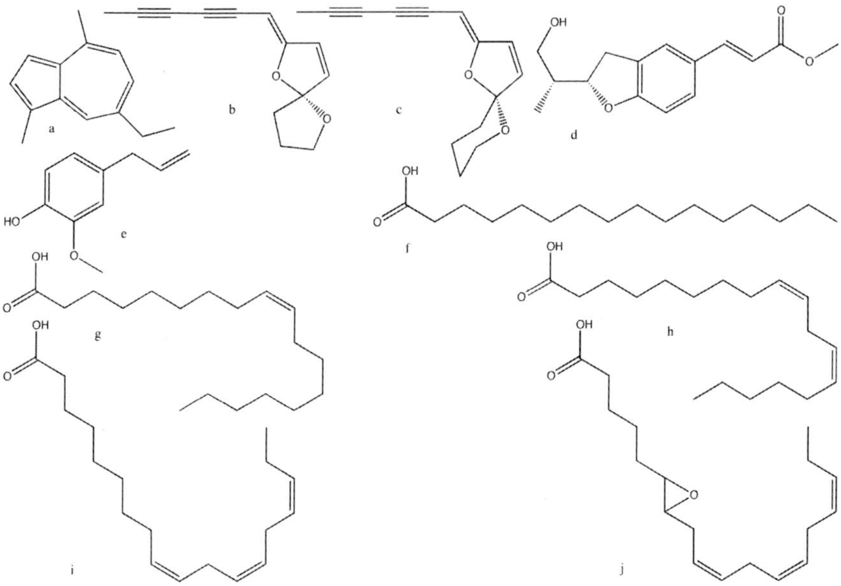

FIGURE 7.3   a) Chamazulene; b) acetylenes (E)-2, c) 3; d) integrinol; e) eugenol; f) palmitic, g) oleic, h) linoleic, i) linolenic, j) 12,13-epoxylinolenic acids.

7.4   DETERMINATION/RISK ASSESSMENT (RA) OF PYRROLIZIDINE ALKALOIDS (PAS) IN *A. CAPILLARIS*

It was informed the simultaneous determination and RA of PAs in *Artemisia capillaris* Thunb. By ultra-high-performance liquid chromatography tandem mass spectrometry (UPLC-MS/MS) together with chemometrics [13]. The PAs are widespread plant secondary metabolites. They are natural toxins because of health risks for livestock and human. *In vivo* and *in vitro* experiments showed that PAs mainly caused damage to the liver; some might harm the lungs *via* the blood vessels and even damage the brain. They can form a series of DNA adducts that cause genotoxicity. The toxicity of PAs is mainly related to chemical structures. The toxic PAs are composed of a 1,2-unsaturated necine base and a necic acid with a branched-chain, whereby the carboxylic ester forms a monoester or a diester group, as does the macrocyclic structure (*cf.* Figure 7.4a). The chemical structures of toxic PAs are classified into retronecine-, otonecine-, and heliotridine-types according to necine bases (Figure 7.4b). The PAs are accumulated primarily in the form of PA N-oxides (PANOs) in many plants.

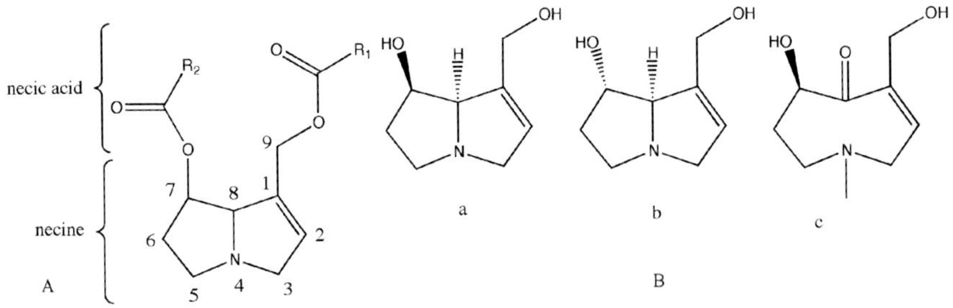

FIGURE 7.4   Toxic PAs: A) basic skeleton; B) types: (i) retronecine; (ii) heliotridine; (iii) otonecine-type.

7.5   SESQUITERPENE LACTONE DERIVATIVES: CYTOTOXIC ACTIVITY AND SELECTIVITY

The STLDs were prepared, and their cytotoxic activity and selectivity of action were studied [14]. In preliminary bioactivity tests, natural STLs, e.g., cumanin (*cf.* Figure 7.5a), showed significant cytotoxic activity. The STLs were used as starting materials for the preparation of a series of oxygenated and oxo-nitrogenous products; the modifications led to an improved activity of the obtained STLDs. The ditriazolyl cumanin STLD was prepared (Figure 7.5b). The incorporation of two triazole groups in cumanin reduced cytotoxicity on normal cells and improved selectivity *vs.* tumor cell lines. The ditriazolyl cumanin STLD was the least toxic *vs.* splenocytes [concentration causing 50% cell death (CC_{50}) = 524.1µM] and presented the greatest selectivity on the tested cell lines. The compound showed a 50% growth inhibition (GI_{50}) of 2.3µM and a selectivity index (SI) of 227.9 on human colon tumor cell line WiDr.

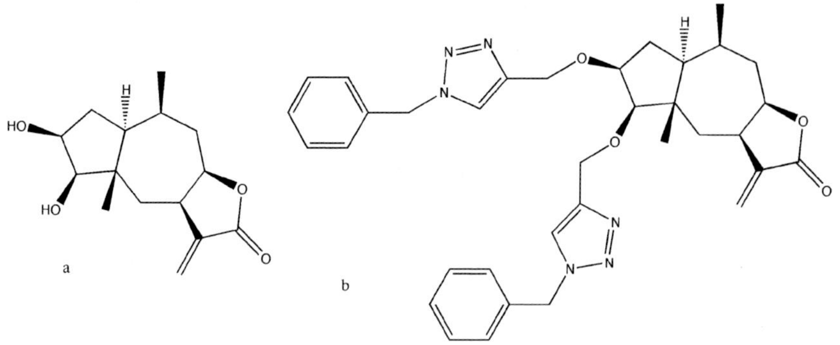

FIGURE 7.5 Structures of (a) natural STL cumanin and (b) ditriazolyl STLD obtained from cumanin.

7.6 FINAL REMARKS

From the present results and discussion, the following final remarks can be drawn:

1. *Artemisia integrifolia* crude lipid extract could be used as a potential treatment to avert hyperlipidaemia. The observed characteristics clearly confirm the medicinal use of *A. integrifolia* in preventing hyperlipidaemia. The lipophilic components and compounds chamazulene, acetylenes (E)-2 and linolenic acid from *A. integrifolia* present positive effects on plasma lipoprotein profiles, suggesting that they show good therapeutic potential in improving hyperlipidaemia. Further studies to confirm the results in other hyperlipidaemia models (e.g., diet-induced obesity) are warranted.

2. Given the results, using natural compounds can be a viable strategy to prepare active molecules. STLs were the starting material for their transformation into several oxygenated and oxo-nitrogenated derivatives by chemical reactions aiming at the hydroxylated positions. The strategy was to obtain derivatives, including functionalities (e.g., acetates, silyl ethers, 1,2,3-triazoles). Although natural STLs showed interesting antiproliferative activity values, a significant number of the synthetic derivatives showed greater activity than the naturally occurring parent product. Many of the synthesized analogs were more selective towards tumor cell lines than normal cells. The ditriazolyl cumanin derivative, proved to be more active

and selective than cumanin in the tested breast, cervix, lung, and colon tumor cell lines. The compound can be considered for further studies and is a possible candidate for developing antitumor agents.

ACKNOWLEDGMENTS

The authors thank support from Generalitat Valenciana (Project No. PROMETEO/2016/094) and Universidad Católica de Valencia *San Vicente Mártir* (Project No. 2019-217-001UCV).

KEYWORDS

- **antiproliferative activity**
- ***Artemisia capillaris* Thunb.**
- ***Artemisia integrifolia***
- **Asteraceae**
- **bioassay-guided fractionation**
- **chemometrics**
- **cumanin**
- **hyperlipidaemia**
- **lipophilic component**
- **pyrrolizidine alkaloid**
- **risk assessment**
- **sesquiterpene lactone**
- **ultra-high-performance liquid chromatography-tandem mass spectrometry**

REFERENCES

1. Guan, X., Ge, D., Li, S., Huang, K., Liu, J., & Li, F., (2019). Chemical composition and antimicrobial activities of *Artemisia argyi* Lévl. et Vant essential oils extracted by simultaneous distillation-extraction, subcritical extraction and hydrodistillation. *Molecules, 24*, Article 483, 1–12.

2. Mighri, H., Hajlaoui, H., Akrout, A., Najjaa, H., & Neffati, M., (2010). Antimicrobial and antioxidant activities of *Artemisia herba-alba* essential oil cultivated in Tunisian arid zone. *C. R. Chimie.*, *13*, 380–386.

3. Wei, X., Xia, L., Ziyayiding, D., Chen, Q., Liu, R., Xu, X., & Li, J., (2019). The extracts of *Artemisia absinthium* L. suppress the growth of hepatocellular carcinoma cells *via* induction of apoptosis *via* endoplasmic reticulum stress and mitochondrial-dependent pathway. *Molecules, 24*, Article 913, 1–14.

4. Castellano, G., Redondo, L., & Torrens, F., (2017). QSAR of natural sesquiterpene lactones as inhibitors of Myb-dependent gene expression. *Curr. Top. Med. Chem.*, *17*, 3256–3268.

5. Torrens, F., & Castellano, G. (in press). Structure-activity relationships of cytotoxic lactones as inhibitors and mechanisms of action. *Curr. Drug Discov. Technol.*

6. Torrens, F., Redondo, L., & Castellano, G., (2017). Artemisinin: Tentative mechanism of action and resistance. *Pharmaceuticals*, *10*, Article 20, 4–4.

7. Torrens, F., Redondo, L., & Castellano, G., (2018). Reflections on artemisinin, proposed molecular mechanism of bioactivity and resistance. In: Haghi, A. K., Balköse, D., & Thomas, S., (eds.), *Applied Physical Chemistry with Multidisciplinary Approaches* (pp. 189–215). Apple Academic-CRC: Waretown, NJ.

8. Torrens, F., & Castellano, G. (2020). Chemical/biological screening approaches to phytopharmaceuticals. In: Pourhashemi, A., Deka, S. C., & Haghi, A. K., (eds.), *Research Methods and Applications in Chemical and Biological Engineering* (pp. 3–12.). Apple Academic-CRC: Waretown, NJ.

9. Torrens, F., & Castellano, G. (2020). Chemical components from Artemisia Austro-Yunnanensis: anti-inflammatory effects and lactones. In: Pogliani, L., Torrens, F., & Haghi, A. K., (eds.), *Molecular Chemistry and Biomolecular Engineering: Integrating Theory and Research with Practice* (pp. 73–83). Apple Academic-CRC: Waretown, NJ.

10. Torrens, F., & Castellano, G. (2020). Triazole-derived, artesunate and metabolic pathways for artemisinin. In: Shinde, R. S., & Haghi, A. K., (eds.), *Modern Green Chemistry and Heterocyclic Compounds: Molecular Design, Synthesis, and Biological Evaluation* (pp. 137–144). Apple Academic-CRC: Waretown, NJ.

11. Cotton, S., (2012). *Every Molecule Tells a Story*. CRC: Boca Raton, FL.

12. Xu, Y., Wang, Q., Bao, W., & Pa, B., (2019). Antihyperlipidemic effect, identification, and isolation of the lipophilic components from *Artemisia integrifolia*. *Molecules*, *24*, Article 725, 1–10.

13. Chen, L. H., Wang, J. C., Guo, Q. L., Qiao, Y., Wang, H. J., Liao, Y. H., Sun, D. A., & Si, J. Y., (2019). Simultaneous determination and risk assessment of pyrrolizidine alkaloids in *Artemisia capillaris* Thunb. by UPLC-MS/MS together with chemometrics. *Molecules*, *24*, Article 1077, 1–17.

14. Beer, M. F., Bivona, A. E., Sánchez, A. A., Cerny, N., Reta, G. F., Martín, V. S., et al., (2019). Preparation of sesquiterpene lactone derivatives: Cytotoxic activity and selectivity of action. *Molecules, 24*, Article 1113, 1–16.

CHAPTER 8

Nanomechanics, Recent Advancements in Nanotechnology, and the Visionary Future

SUKANCHAN PALIT

43, Judges Bagan, Post-Office-Haridevpur, Kolkata–700082, India,
Tel.: 0091-8958728093, E-mails: sukanchan68@gmail.com,
sukanchanp@rediffmail.com, sukanchan92@gmail.com

ABSTRACT

Science, technology, and engineering are in the avenues of immense scientific hardship and scientific introspection. The status of the science of nanotechnology is today groundbreaking and immensely inspiring. Nanomechanics are the marvel of science and engineering today. Investigation of nanomaterials (NMs) and engineered NMs has urged scientists and engineers to delve deep into the world of nanomechanics. This is a new generation science integrating mechanical engineering and nanotechnology. The author in this paper deeply unraveled the scientific truth and the scientific regeneration of both nanotechnology and nanomechanics. This treatise broadly investigates the recent developments in the field of nanotechnology and nanomechanics. The vast scientific vision, the scientific subtleties, and the futuristic vision of mechanical engineering will bring a new era in the field of nanomechanics. Nanomechanics are a domain which integrates diverse areas of science and engineering. Applied physics and electronics engineering are other areas of scientific endeavor in nanomechanics today. Mankind's immense scientific prowess and determination, the world of scientific validation and the need for nanotechnology in human society will all be the torchbearers towards a new era in science and technology. Today in the global scientific scenario, there should be scientific and technological validation of every areas of scientific endeavor. In the field of nanomechanics, a similar vision is needed since it is a new branch of research pursuit. The diverse areas of nanotechnology are dealt with immense adroitness in this chapter.

8.1 INTRODUCTION

The domain of science of nanotechnology is undergoing drastic changes. Similarly, environmental engineering and chemical process engineering are in the path of newer scientific divination. Nanomechanics are a branch of nanotechnology whose research directions are inspiring and thought-provoking as mankind treads forward. Global climate change, depletion of fossil fuel resources and frequent environmental disasters has deeply urged the world of science and technology to move towards newer paths of scientific endeavor and innovations. In this chapter, the author with immense scientific ingenuity deals with also recent scientific developments in nanotechnology and nanomechanics. The needs of human civilization are the applications of renewable energy technology. Here also there are lots of applications of nanotechnology. Rapid industrialization and vast scientific progress has devastated the global environmental scenario. Lack of pure drinking water is a monstrous issue in many highly developed and developing countries around the world. Heavy metal and arsenic ground-water contamination is a global calamity and disaster of immense scientific concern. The world also stands beleaguered at the monstrous issue of depletion of fossil fuel resources. Here comes the veritable concerns and importance of scientific innovations and discoveries. Energy engineering and energy sustainability are the needs of humanity today. The world stands deeply strangled with ever-growing concerns of non-renewable energy. The success of this treatise lies at the hands of scientists and engineers globally. The vision of nanotechnology will surely open new thoughts and newer scientific advancements in diverse areas of science and engineering globally. This treatise is an eye-opener towards newer developments in nanotechnology and nanomechanics.

8.2 THE VISION OF THIS STUDY

Scientific advancements in the field of nanomechanics are latent yet ground-breaking. Nanotechnology is a marvel of science and engineering today [11–14]. The vision of this chapter is to unravel the scientific ingenuity and the scientific truth behind application of nanomechanics in the pursuit of science and engineering. The areas investigated in this chapter are nano-materials (NMs) and engineered NMs. Technology and engineering of NMs and engineered NMs are moving very fast overcoming one scientific boundary over other. The aim and objective of this study is to delineate and elucidate the basics and fundamentals of NMs and engineered NMs and the vast domain of nanomechanics. Human civilization and scientific advancements today are in the path of immense retrogression and deep reformation as environmental engineering and environmental protection stands firm in the midst of scientific contemplation. The targets of this study are:

- A scientific understanding on the field of nanomechanics.
- A deep investigation into the field of application of nanotechnology in environmental protection.
- Technological advancements and scientific verve in the field of NMs and engineered NMs.
- Scientific validation in the recent advancements of nanotechnology and nano-engineering.
- A vast scientific stride in the field of environmental engineering, chemical engineering and diverse areas of engineering and science [11–14].

Humanity, science, and engineering are moving and treading in the path of new reformation and a new beginning in this century. The world of mechanical engineering and mechanics are today replete with immense scientific and academic rigor. The challenges and the deep vision are related in this paper. A deep scientific and engineering investigation in the field of nanomechanics stands as a cornerstone of this well researched treatise. Today the domain of environmental engineering is highly challenged and immediate action is needed in the mitigation of climate change and environmental calamities. Thus the need of a comprehensive paper [11–14].

8.3 THE SCIENCE OF SUSTAINABILITY AND THE VISION FOR THE FUTURE

Sustainable development whether it is energy or environmental is the utmost necessity of human civilization's progress today. The inspiration of scientific advancements today lies in the domain of nanotechnology, nano-engineering, environmental engineering, water resource management, groundwater remediation and industrial wastewater management. Human scientific imagination is at its helm as world faces immense scientific issues such as climate change, loss of ecosystem biodiversity and loss of fossil fuel resources. In this paper, the author puts forward to the scientific platform the needs of human civilization which are nanotechnology and nano-engineering. The science of sustainability is in the path of new scientific truth and a deep awakening. The vision for the future in the field of mechanical engineering and chemical engineering are in the equal sense in the path of scientific truth and scientific ingenuity. With rapid industrialization and the burden of groundwater heavy metal contamination, the world of science and technology stands flummoxed and thus needs to be reframed with the progress of human civilization. Sustainability whether it is economic or social will pave the way towards a new scientific revolution. Diverse branches of science and engineering such as mechanical engineering and chemical engineering will thus enter into a new phase of regeneration, vision, and might [11–14].

8.4   WHAT DO YOU MEAN BY NANOMECHANICS?

Nanomechanics is a branch of nanoscience studying fundamental mechanical (elastic, thermal, and kinetic) properties of physical and engineering systems at the nanometer scale. Technological vision and scientific motivation, the scientific candor and the futuristic vision of mechanics and mechanical engineering will transform the entire scientific genre and civilization. Nanomechanics have emerged on the crossroads of classical mechanics, solid-state physics, statistical mechanics, material science, composite science and quantum physics and quantum chemistry. Today nanomechanics provides a strong foundation of nanotechnology. The vast challenges of science and engineering need to be reframed globally as regards application of nanotechnology and nanomechanics in human society. Nanomechanics is a branch of nanoscience which deals with the study of and applications of fundamental mechanical properties of physical and engineering systems at the veritable nanoscale such as elastic, thermal, and kinetic material properties. Often nanomechanics are regarded as a branch of nanotechnology, i.e., an applied area with a deep focus on the mechanical properties of engineered nanostructures and nanosystems (systems with nanoscale components of importance). Examples of the latter include nanoparticles (NPs), nanopowders, nanowires, nanorods, nanoribbons, nanotubes, including carbon nanotubes (CNTs) and boron nitride nanotubes (BNNTs), nanoshells, nanomembranes, nanocoatings, nanocomposite/nanostructured materials, nanomotors, etc. Validation of science and engineering in the field of nanotechnology is opening new doors of innovation and instinct in human scientific progress in years to come [11–14, 25–27].

8.5 RECENT SCIENTIFIC PURSUIT IN THE FIELD OF NANOTECHNOLOGY

Nanotechnology research directions are moving towards new scientific revamping. The intricacies, the challenges, and the vision of nanotechnology research are in the path of sound engineering contemplation. The struggle and the vision of nanotechnology and nanomechanics thus need to be revamped with the human scientific progress. In this section, the author elucidates the success and the vision of recent developments in nanotechnology.

Davies [1] gave a detailed oversight of next-generation nanotechnology. This treatise is a Woodrow Wilson International Center for Scholars Report. The author elucidated on the future of nanotechnology, existing oversight, and next-generation nanotechnology and the future of oversight. Scientific verve and motivation and technological embellishment are the needs of scientific research pursuit in nanotechnology [1]. Nanotechnology is the true achievement of human civilization today. Since 1980s, the scientific capability of United States federal agencies responsible for environmental health and chemical process safety has steadily eroded [1]. The agencies, the scientists and engineers cannot perform their basic functions now, and they are unable to cope up with the new challenges and the vision of the 21st century [1]. This paper reviews some of the challenges, focusing on next-generation nanotechnologies and nano-science and suggests changes that could revitalize the health and safety agencies. Oversight of new technologies in this century will surely occur in a context characterized by rapid scientific advancement, accelerated application of nanotechnology and nano-engineering and frequent product changes [1]. Technological verve and scientific embellishment need to be reorganized and revitalized and this report gives a wider glimpse towards these issues. The products in this century will be technically complex, pose potential health and environmental problems and have a sound effect on many sectors of society simultaneously [1]. The federal regulatory bodies already suffer from under-funding and bureaucratic challenges, but they will require more than just increased funding and minor rule changes to deal with potential adverse effects of the new technologies. To stimulate discussion, this treatise outlines a new federal Department of Environmental and Consumer Protection [1]. The new agency will have three directions: oversight, research, and assessment and monitoring. A new integrated approach to industrial and domestic pollution control was necessary even before the Environmental Protection Agency was created in 1970 and since then the need has increased. The design of the proposed new scientific agency incorporates the proposals for an Earth System Agency and a Bureau of Environmental Statistics [1]. The paper describes some of the recent developments that will determine the future of technology and some changes that would equip the federal government to deal with the new 21st-century science and technology. The drastic changes and the challenges of new technologies such as nanotechnology are described in minute details in this report [1]. The novel characteristics of NMs mean that risk assessments (RA) developed from ordinary materials may be of limited use in determining the risks and the environmental health of the products of nanotechnology [1]. A growing body of evidence points to the potential for unusual health and environmental risks of NMs and engineered NMs. NMs have a much larger ratio of surface area to mass than ordinary materials do. Oversight consists of obtaining risk information and acting on it to prevent health and environmental damage. An underlying importance of this paper is the oversight on nanotechnology to prevent damage and the proliferation of the science of nanotechnology [1]. The United States and Europe are at the forefront of the necessity of oversight and regulation is necessary for the proper absorption of nanotechnology in human civilization's progress [1].

European Commission Report [2] deeply investigated with insight and purpose successful European nanotechnology research. The report also elucidates outstanding science and technology to match the needs of the future society [2]. Nanotechnology and nano-vision are the hallmarks of human scientific regeneration today. Nanotechnology is the new frontier of science and technology in the global scenario working at the scale of individual molecules. There are diverse research directions in the field of nanotechnology today [2]. The impact of nanotechnology on human society is the ultimate hallmark of scientific endeavor today. In the European Union, research on nanotechnology and nano-engineering takes a very special place to the extent that information, communication, and fostering a debate on nanotechnology research has already become an essential part of many diverse European policy initiatives and global research and development initiatives [2]. Human civilization, human scientific genre and vast scientific motivation are the scientific embellishment globally [2]. Communicating nanotechnology research to the global scientific arena is a necessity of every nation around the globe. The vision of nanotechnology is surely splendid and awe-inspiring to every citizen around the global. This report depicts profoundly the appropriate methodologies of nanotechnology applications to citizens in the European Union and nations around the world [2]. Technological fervor, scientific truthfulness and the sagacity of human scientific pursuit will all be the forerunners towards a new awakening in science and engineering globally [2]. Nanomedicine with its diagnostics, drug delivery, and regenerative medicine, nanotechnology for environment/energy and electronics/information and communication technology are the cornerstones of scientific research pursuit globally today. The chapters in human scientific progress and human civilization are today wide open as science treads forward towards a new era. This report widely opens new arenas of research pursuit in the domains of nano-science, nanotechnology, and nano-engineering [2].

Federal Institute of Occupational Safety and Health, Germany Report [3] discussed with immense scientific profundity health and environmental risks of NMs and nanotechnology and its vast research strategy [3]. This report elucidates on strategic aims, research, and work areas, improvement in the comparability of studies on toxicology, ecotoxicology, environmental behavior of NMs, and characterization of their physico-chemical properties, determination of the chemical reactivity, vast areas of occupational risk management, and the areas of occupational health. Today is a highly challenged technology driven society [3]. The world stands really and veritably flummoxed at the global state of environment and ecosystem. Human civilization has to garner enough might and vision in true realization of nanotechnology, NMs, and diverse areas of engineering and science. Environmental protection and groundwater heavy metal contamination are today in a situation of scientific jinx and immense scientific travails [3]. The contribution of scientists and engineers should be towards greater emancipation of environmental remediation and greater amelioration of climate change. As an important future technology, nanotechnology presents an enviable opportunity for positive economic development in the global scenario today [3]. Every nation around the world are today at its might and vision in true realization of nanotechnology in human progress [3]. According to the present knowledge, the insoluble and poorly soluble NMs are of particular toxicological scientific and engineering relevance. Here comes the need of a detailed investigation of effects of NMs on human health and the environmental health of global eco-system [3]. Chemicals legislation does not provide for a specific procedure for testing and assessment of NMs and engineered NMs. Since exposure of humans and environment, the toxicological properties and risks cannot yet be evaluated, the need to conduct detailed investigations and close scientific gaps in knowledge by research and assessment activities is highly required. As a developed country, Germany is at the forefront of immense revamping in the application of nanotechnology to human society. The report delineates the following strategic aims in the field of nanotechnology which are:

- Risk oriented approach [3].
- Comprehensive RA and deep risk characterizations.
- Integration into the statutory regulatory framework [3].
- Technological vision and vast scientific motivation in the field of NMs applications.
- Research that is highly application oriented [3].
- Sustainability principles in the application of nanotechnology in human society.
- More efficient structures in the targeted promotion of research.
- A sound technological and engineering support in nanotechnology realization.
- Transparency and vast public discourse [3].

Assessment of the novelty of NMs is one of the hallmarks of the report. A deep elucidation on the fate of NMs in the environment is the cornerstone of this report.

Ursino et al. [4] discussed with lucid and cogent scientific conscience progress of nanocomposite membranes for water treatment. Technological vision and vast scientific profundity and motivation are the needs of application of nanotechnology in human society today [4]. The use of membrane based technologies has been applied with immense vision and seriousness for water treatment applications. The limitations of conventional polymeric membranes have led to the addition of inorganic fillers in improving their performance [4]. In recent years, nanocomposite membranes have greatly attracted scientists, engineers, and technologists for water treatment applications such as wastewater treatment, water purification, drinking water treatment, removal of microorganisms, chemical compounds, heavy metals etc. Polymeric membranes are widely used for water and wastewater treatment, e.g., waste streams from agro-food, textiles, and petroleum industries or removal of pollutants from drinking water, enabling the concentrate to be treated or discharged and thereby reducing the contaminants discharged into wastewater [4]. Pressure driven membrane processes such as Microfiltration, Ultrafiltration, Nanofiltration, and Reverse Osmosis are considered as highly promising alternatives for the removal of large amounts of organic micropollutants [4]. Here comes the importance of a detailed investigation of nanocomposite membranes and its applications. The authors in this paper also delineated the applications of various NMs in water and wastewater treatment. The applications of CNTs, zinc oxide, graphene (GR) oxide, 2-dimensional materials, and some other novel nano-sized materials are the other cornerstones of this well researched paper [4].

Uddin et al. [5] discussed with vast scientific far-sightedness application of NMs in the remediation of textile effluents from aqueous solutions. Technological verve and motivation, the deep scientific ingenuity and vision will today surely unravel the scientific intricacies of environmental remediation globally [5]. Textile dyes, if present in wastewater, have hazardous and deleterious effects on the life of aquatic animals and human beings [5]. The challenges and the targets of wastewater treatment are immense and groundbreaking today. Textile, finishing, and dye manufacturing industries release a large quantity of wastewater containing toxic dye-stuffs into the aquatic systems. Dyes are widely used for coloring products in several industries such as textiles, leather, paper, rubbers, paint, tannery, pharmaceuticals, plastics, foodstuffs, cosmetics, etc. [5]. This paper also classifies various types of dyes. Adsorption of various dyes on NMs stands as a major hallmark of this research pursuit. The applications of various types of NMs and engineered NMs in water and wastewater treatment are the major pillars of this paper [5].

Darwish et al. [6] discussed and described with vast scientific vision functionalized nanomaterial for environmental techniques. The world of scientific challenges and technological ingenuity in the field of nanotechnology and environmental remediation are deeply investigated in this paper [5]. Nanotechnology is today envisioning the development of new solutions to environmental problems due to high surface area and associated high reactivity on a scale ranging from one to a few hundred nanometers that are surely not absorbed in the macroscale. The most attractive nanomaterial for environmental remediation are derived from silica, noble metals, semiconductors, metal oxides, polymers, and carbonaceous materials [5]. The categories of wastewater and drinking water techniques include adsorption, photocatalytic degradation and disinfection, nanofiltration, and monitoring of inorganic and organic pollutants [5]. The authors discuss nanoadsorption, membranes, and membrane processes, nanophotocatalysis, and nanosensing in detail. Scientific and technological abundance and ingenuity are the hallmarks of this chapter [5]. Limitations of NMs used for environmental techniques are the other cornerstones of this paper. Methods of NMs' functionalization are deeply investigated in this treatise [5]. Applications of functionalized silica-based NMs and carbonaceous NMs are the other challenges of this paper. The entire areas of nanotechnology developments for environmental applications are dealt with insight and vision in this paper. The vision of science and engineering today needs to be reframed as civilization treads forward [5]. In a similar manner, functionalized NMs and its applications stand today as a major scientific understanding globally. These areas of immense importance and vision are delineated in this chapter [5].

Human scientific vision in nanotechnology and nanomechanics are in the avenue of newer regeneration. The world of science and engineering is in the midst of immense scientific intricacies and profundity. Diverse areas of science and technology are in the process of new reorganization. Nanotechnology in the similar manner needs to be reorganized as regards its applications, prospects, and challenges. This chapter opens up new thoughts and new ingenuity in engineering science in years to come.

8.6 RECENT SCIENTIFIC ENDEAVOR IN THE FIELD OF NANOMECHANICS

Nanomechanics today are a recent marvel of science. It is an integration of nanotechnology and mechanical engineering. The world of nanotechnology is poised towards a newer beginning. In this paper, the authors with deep scientific divination uncover the scientific intricacies in the field of nanomechanics and a deep investigation into recent developments in nanotechnology. Vast scientific imagination needs to be thoroughly re-envisioned as regards application of nanotechnology in environmental engineering and other diverse areas of science and engineering. The author pinpoints the success of nanomechanics in solving difficult engineering problems. The world of engineering science is thus highly challenged as mechanical engineering surges forward towards a new visionary scientific era.

Porok et al. [6] discussed and deliberated on the science of micro- and nano-mechanics. It has been known for some time that materials and structures with small dimensions do not behave in the same manner as their bulk counterparts [6]. Micro- and nano-mechanics stands in the midst of deep scientific redeeming and vast scientific ingenuity. The aspect that NMs behave in a different manner was first observed in thin films (TFs) where certain defect structures were found to have deleterious effects on the film's structural integrity. This is of vast importance since TFs are employed as components in microelectronics and microelectromechanical systems [6]. The physics of nanomechanics today is in the vistas of sound learning and vast scientific profundity. In this well researched treatise, the authors discussed with vision and engineering conscience TFs, TF measurement techniques, testing methods based on microelectromechanical system technology, nanoscale measurement techniques, and measurement techniques for nanotubes and nanowires [6]. Other areas discussed in this paper are frontiers in nanoscale experimental techniques. Theoretical modeling and scaling are the other cornerstones of this paper. Silicon nanowires and its structure and energetic stands as the focal points of research endeavor in this paper [6].

Liu et al. [7] discussed with vast scientific ingenuity and farsightedness computational nanomechanics and materials. Today many areas of research are progressing at a rapid pace due to the combined effect of science, technology, and engineering [7]. In some of the cases, fields of research that are stagnating under the exclusive domain of one discipline have been involved with new discoveries through collaborations with scientists, engineers, and practitioners of the second discipline. In computational mechanics, the particular concern is about the technological engineering interest by the combination of engineering sciences, engineering technology, and basic sciences through modeling and simulation [7]. These scientific goals resulted in the emergence of nanotechnology and the related forays by nanoscale researchers [7]. Human scientific advancement and engineering profundity today in the critical juncture vast scientific reemergence and redeeming [7]. Nanotechnology today has changed the human society totally. In this paper the authors summarizes the strengths and limitations of currently available multiple scale techniques, where the emphasis is on bridging scale method, multi-scale boundary conditions, and multi-scale fluidics. The rapid advances in nanotechnology, nanomechanics, and NMs offer vast scientific potentials in defense, homeland security, and private industry [7]. A deep emphasis on nanoscale entities will make manufacturing technologies and infrastructure more sustainable in terms of energy usage and environmental pollution control. With the confluence of the science of nanotechnology, the availability of experimental and scientific tools to synthesize and characterize systems in the nanometer scale and the computational tools widely accessible to model micro-mechanical systems are investigated in detail in this chapter [7]. The authors in this paper discussed molecular dynamics, and the vast intriguing area of computational mechanics. The world of science and engineering today is abounding with vision, scientific forbearance, and immense insight. This paper opens up new thoughts and new research directions in the field of nanomechanics, computational mechanics, and the holistic domain of mechanical engineering. Thus, a new beginning will usher in and the world of science and engineering will tread forward [7].

Backes et al. [8] discussed with immense scientific insight and vast lucidity nanomechanics and nanorheology of microgels at the interfaces. This review addresses nanomechanics and nanorheology of stimuli responsive microgels adsorbed at an interface [8]. Technological and engineering profundity and innovation are the hallmarks of research pursuit in mechanical engineering and polymer science today. Nanomechanics are today the integration of physics, nanotechnology, and mechanical engineering. The vision, the vast challenges, and the engineering profundity will truly unravel the success of science of nanomechanics today [8]. Rheology and nanomechanics are the vision of tomorrow in its application, might, and vision. The focus of this chapter is on the correlation between the swelling abilities of the gels and their mechanical properties [8]. Hydrogels are defined as cross-linked polymeric networks that are able to veritably swell in water [8]. In this review, mainly chemical cross-linked hydrogels will be considered, which show a volume phase transition triggered by temperature rise [8]. Besides macrogels, in the last 20 years microgels got more and more impact. The advantages of microgels are the faster response kinetics due to their smaller distances [8]. In this paper, the author deeply discusses atomic force microscopy applications, static force measurements, dynamic force measurements, and the vast and intuitive areas of nanomechanics and nanorheology. The other areas of scientific ingenuity of this paper are the domains of applications of nanomechanics and nanorheology [8]. This review addresses the mechanical properties of microgels adsorbed at interfaces [8]. The authors also investigated the correlation between the swelling behavior and the mechanical properties of the microgels [8]. This review shows that there are still open research questions about nanomechanics and nanorheology of microgels and it is really envisioned that it will create a sound impact in its vast and varied applications [8].

Sheikh et al. [9] in a concise review dealt with recent advances in nanotechnology and potential prospects in neuromedicine and neurosurgery. The advent and popularity of minimal access surgeries and micro-medicine tools incur the current forays in the utilization of nanotechnology and nanodevices (NDs) in medical fields [9]. Technological abundance and resurgence are the needs of a truly new era in the field of NDs today. The applications of nanotechnology in the field of neurosurgery and neuroscience need to be vehemently understandable for more transparent solutions and vast innovations [9]. This review examines the recent developments and potentials of the vast utilization of nanotechnology in nanomedicine and neurosurgery [9]. The application of nanotechnology in the medical field resulted in a new frontier in medicine described as nanomedicine which is the application of structured NPs such as dendrimers, carbon fullerenes, and nanoshells [9]. The futuristic vision of nanomedicine will surely open new vistas of scientific profundity and deep engineering ingenuity in science and engineering globally [9]. The authors discussed in detail nanotechnology in neuro-drug delivery, functional NPs for diagnostic imaging, NDs in neurological surgery procedures, and a vast scientific understanding and ingenuity in the field of nanomedicine. Nanotechnology is today providing a future platform for further research and development in neuroscience and neurosurgery. These areas of scientific research pursuit are dealt with immense might and vision in this paper [9].

Zhang et al. [10] described and delineated with cogent insight recent advances in nanotechnology applied to biosensors. In recent years, there has been an unparalleled advance in the application of NMs in biosensors [10].

Here the authors deeply reviewed NMs such as gold NPs, CNTs, magnetic NPs, and quantum dots [10]. Their application in biosensors is the focal point in this chapter. A new interdisciplinary field of biological detention and material science are the cornerstones of this research endeavor today [10]. A biosensor is a device incorporating a biological sensing element connected to a transducer. In recent years, with the development of nanotechnology, a lot of novel NMs are being fabricated. They all have vast and varied applications [10]. Here the authors reviewed with insight some of the main advances in this field for the past few years, vastly explore the application prospects and deeply discuss the issues, approaches, and the challenges in NMs based biosensor technology [10]. NMs can improve mechanical, electrical, optical, and magnetic properties of biosensors. Nanomaterial based biosensors show great attractive prospects, and will find applications in clinical diagnosis, food analysis, process control, and environmental remediation [10].

The success of nanotechnology and nanomechanics are changing the global scenario of science and engineering. Research directions needs to be re-envisioned and deeply re-envisaged in the field of nanotechnology and nano-engineering. In this chapter, the authors elucidate on the futuristic vision of both nanotechnology and nanomechanics [11–14].

8.7 TECHNOLOGICAL ADVANCEMENTS IN NANOTECHNOLOGY AND THE VAST VISION FOR THE FUTURE

Technological divination, the vast scientific ardor, and the utmost needs of the human civilization are veritably opening up new doors of scientific intuition in the field of nanotechnology today. Human challenges in the field of water science and water purification are enormous, vast, and varied. The needs of human civilization are the domains of energy and environmental sustainability. The needs of renewable energy and reliability engineering are of immense importance as science and engineering moves forward. Sustainable development whether it is social, economic, energy or environmental is the needs of human civilization today. Human factor engineering, integrated water resource management and wastewater management are equally of immense importance today. The status and scientific stance of human civilization are immensely grave and mind-boggling to scientists and engineers around the globe. Successive human generations around the world needs to be aroused at the ever-growing concerns of climate change and loss of ecological biodiversity. Thus, novel technologies such as nanotechnology need to be reframed and re-organized with the surge of human civilization. This chapter opens up new thoughts and newer scientific imagination in the field of nanotechnology and its applications. Today is highly technology-driven society. The perspectives of science and engineering needs to be re-envisioned as regards application of nanotechnology in human society. The challenges and the vision of energy and environmental sustainability need to go a long way in the true emancipation of science and engineering globally today. Technology management and reliability engineering are the utmost necessities of human scientific endeavor today [15–20].

8.8 MODERN SCIENCE, NANOTECHNOLOGY, AND A VISION FOR THE FUTURE

Modern science and nanotechnology are today the necessities of science and engineering globally today. Nanotechnology has applications today in all most every branch of science and technology. The vision for the future needs to be re-envisioned and re-organized as nanotechnology moves forward. Today is the scientific world of technology management, project management, and reliability engineering. Water science and technology and water purification are veritably linked with nanotechnology today. NMs and engineered NMs are used as adsorbents in water purification. The vision and the challenges of science are enormous as regards the applications of nanotechnology in water and wastewater treatment. CNTs are presently doing scientific wonders as regards its applications in diverse areas of science and engineering. Nanotechnology is also linked with integrated water resource management and industrial ecology in industrialized and developed nations around the world. Modern science and the sagacity of science and technology needs to be re-envisioned and re-organized as regards application of nanotechnology and nanomechanics in vast and varied areas of science and technology. The world of science and technology today stands immensely mesmerized with the growing concerns of heavy metal and arsenic groundwater and drinking water contamination. Here comes the importance of nanotechnology, nano-engineering, reliability engineering, technology management, and the vast world of project management. Integrated water resource management is today creating wonders in developing and poor nations around the world. Thus the need of watershed management or integrated water resource management and waste management [21–27].

8.9 FUTURE SCIENTIFIC RECOMMENDATIONS IN THE FIELD OF NANOTECHNOLOGY AND NANOMECHANICS

Nanotechnology and nanomechanics are the needs of science and engineering today. Mechanical engineering frontiers are in the path of a newer beginning. Scientific vision in the field of both mechanical engineering and nanotechnology are today surpassing vast and versatile frontiers. The needs of human advancement are environmental engineering, chemical process engineering, and the vast domain of renewable energy and energy engineering [25–27]. Environmental remediation and nanotechnology advances should go parallel in scientific research directions. The scientific potential for nanomechanics is immense and far-reaching today. Scientific forbearance, scientific ardor, and deep scientific divination are the veritable pallbearers towards a new scientific generation in nanotechnology and other diverse areas of engineering such as chemical process engineering and environmental engineering. Technology redeeming is veritably the need of human mankind today. These are the areas which need to be targeted as man moves forward. Energy and environmental sustainability are the marvels of scientific research pursuit today. Today the scientific world stands mesmerized with the advancements in space technology [25–27]. Nanotechnology, NMs, and nanomechanics does not stand behind. There are absolutely exemplary achievements in nanotechnology and nano-engineering today. The author tried to elucidate with might and vision the success of nanomechanics in human scientific advancements. The intricacies of application of NMs and engineered NMs are the focal points of this paper.

8.10 FUTURE FLOW OF SCIENTIFIC THOUGHTS IN THE FIELD OF NANOMECHANICS

Nanotechnology and nanomechanics are today the pivotal point of research endeavor globally today. Future flow of scientific thoughts should be directed towards applications of nanomechanics in diverse areas of science and engineering. Sustainable development and engineering science are two opposite sides of the visionary scientific coin today. Environmental and energy sustainability are the pillars of scientific endeavor today. In the global scientific scenario, environmental catastrophes, climate change, and depletion of fossil fuel sources are challenging the scientific firmament. Technological and scientific validation, the futuristic vision of environmental sustainability, and the basic needs of human society will all usher in a new era in environmental engineering. Provision of clean drinking water is the need of human society today. Thus, future research directions should be directed towards achievement of energy and environmental sustainability. Scientific thoughts in the field of nanotechnology needs to be streamlined and reorganized as the world stands devastated with environmental catastrophes and fossil fuel issues. Nanomechanics and the world of nanotechnology are in the crucial juncture of scientific vision and contemplation. Industrialized nations around the globe are in the midst of a deep crisis such as mitigation of climate change and the world of challenges in industrial wastewater treatment.

8.11 CONCLUSION AND SCIENTIFIC PERSPECTIVES

Humanity and human civilization are today in the midst of deep scientific vision and contemplation. Scientific perspectives in the field of nanotechnology need to be re-envisioned with the march of environmental engineering and chemical process engineering. Environmental remediation, drinking water treatment, and industrial wastewater treatment are in the process of a new beginning and a global revival. Scientific divination and deep scientific redeeming are today the necessities of science, technology, and engineering globally. Water science and water purification are today destroying the scientific landscape. Thus the need of nanotechnology and applications of nanomechanics. Overall, nanotechnology research directions should be targeted towards more scientific emancipation in the field of NMs and engineered NMs. The burning question of arsenic and heavy metal contamination in groundwater needs to be answered with immediate effect. Here come the need of participation of civil society, scientists, and engineers. Today the world is a technology driven society. Science and technology is marching forward towards a newer age of immense might and vision. Nanotechnology is a veritable lantern of human civilization today. The scientific perspectives of nano-science, nanotechnology, and nano-engineering needs to be re-envisioned and redrawn with the progress of human mankind today. The challenges, the vision, and the targets of nanotechnology need also to be reorganized with the march of humanity forward. Technology frontiers globally are highly retrogressive today. This challenge and this task should be shouldered by scientists and engineers. The global scientific questions of water purification, drinking water treatment and groundwater contamination still remain unanswered. The scientific rationale behind water purification thus needs to be re-organized and researched with immense vision. Thus, a newer eon in the field of nanotechnology will usher in and a new global order in the field of chemical engineering and environmental engineering will be envisioned. Then technological barriers will be ameliorated and a newer visionary global order will emerge.

KEYWORDS

- **boron nitride nanotubes**
- **carbon nanotubes**
- **environmental engineering**
- **nanomaterials**
- **nanoribbons**
- **nanotechnology**
- **nanotubes**

REFERENCES

1. Davies, J. C., (2009). *Oversight of Next Generation Nanotechnology*. Woodrow Wilson International Center for Scholars, (Report), (Project of Emerging Technologies).
2. European Commission Report, (2011). *Successful European Nanotechnology Research, Outstanding Science and Technology to Match the Needs of Future Society*. Edited by the European Commission, Directorate-General for Research and Innovation, Directorate Industrial Technologies, Luxembourg.
3. Federal Institute for Occupational Safety and Health, (2007). Germany report. *Nanotechnology: Health and Environmental Risks of Nanomaterials*. Research Strategy.

4. Ursino, C., Castro-Munoz, R., Drioli, E., Gzara, L., Albeirutty, M. H., & Figoli, A., (2018). Progress of nanocomposite membranes for water treatment. *Membranes, 8, 18*, 1–40.

5. Uddin, M. K., & Rehman, Z., (2018). Applications of nanomaterials in the remediation of textile effluents from aqueous solutions. In: Shahid-Ul-Islam, & Butola, B. S., (eds.), *Nanomaterials in the Wet Processing of Textiles* (pp. 135–161). Scrivener Publishing.

6. Darwish, M., & Mohammadi, A., (2018). Functionalized nanomaterial for environmental techniques, Chapter-10. In: Chaudhery, M. H., & Ajay, K. M., (eds.), *Nanotechnology in Environmental Science* (pp. 315–349). Wiley-VCH Verlag GMBH & Co-Germany.

7. Porok, B. C., Zhu, Y., Espinosa, H. D., Guo, Z., Bazant, Z. P., Zhao, Y., & Yakobsen, B. I., (2004). Micro- and nanomechanics. In: Nalwa, H. S., (ed.), *Encyclopedia of Nanoscience and Nanotechnology* (pp. 555–600). American Scientific Publishers, USA.

8. Liu, W. K., Karpov, E. G., Zhang, S., & Park, H. S., (2004). An introduction to computational nanomechanics and materials. *Computational Methods in Applied Mechanical Engineering, 193*, 1529–1578.

9. Backes, S., & Von, K. R., (2018). Nanomechanics and nanorheology of microgels at interfaces. *Polymers, 10*(978), 1–23.

10. Sheikh, B. Y., (2014). *Recent Advances in Nanotechnology: Potential Prospects in Neuromedicine and Neurosurgery.* Nanoscience & technology, Open access, www.symbiosisonlinepublishing.com (accessed on 22 February 2020).

11. Zhang, X., Guo, Q., & Cui, D., (2009). Recent advances in nanotechnology applied to biosensors. *Sensors, 9*, 1033–1053. doi: 10.3390/s90201033.

12. Palit, S., (2017). Application of nanotechnology, nanofiltration, and drinking and wastewater treatment: A vision for the future. Chapter-17. In: Alexandru, M. G., (ed.), *Water Purification* (pp. 587–620). Academic Press, USA.

13. Palit, S., (2016). Nanofiltration and ultrafiltration—the next generation environmental engineering tool and a vision for the future. *International Journal of Chem. Tech. Research, 9*(5), 848–856.

14. Palit, S., (2016). Filtration: Frontiers of the engineering and science of nanofiltration: A far-reaching review. In: Ubaldo, O. M., Kharissova, O. V., & Kharisov, B. I., (eds.), *CRC Concise Encyclopedia of Nanotechnology* (pp. 205–214). Taylor and Francis.

15. Palit, S., (2017). Advanced environmental engineering separation processes, environmental analysis and application of nanotechnology: A far-reaching review. Chapter-14. In: Hussain, C. M., & Kharisov, B., (eds.), *Advanced Environmental Analysis: Application of Nanomaterials* (Vol. 1, pp. 377–416). The Royal Society of Chemistry, Cambridge, United Kingdom.

16. Hussain, C. M., & Kharisov, B., (2017). *Advanced Environmental Analysis: Application of Nanomaterials* (Vol. 1). The Royal Society of Chemistry, Cambridge, United Kingdom.

17. Hussain, C. M., (2017). Magnetic nanomaterials for environmental analysis. Chapter-19. In: Hussain, C. M., & Kharisov, B., (eds.), *Advanced Environmental Analysis: Application of Nanomaterials* (Vol. 1, pp. 3–13). The Royal Society of Chemistry, Cambridge, United Kingdom.

18. Hussain, C. M., (2018). *Handbook of Nanomaterials for Industrial Applications.* Elsevier, Amsterdam, Netherlands.

19. Palit, S., & Hussain, C. M., (2018). Environmental management and sustainable development: A vision for the future. In: Chaudhery, M. H., (ed.), *Handbook of Environmental Materials Management* (pp. 1–17). Springer Nature Switzerland A.G.
20. Palit, S., & Hussain, C. M., (2018). Nanomembranes for environment. In: Chaudhery, M. H., (ed.), *Handbook of Environmental Materials Management* (pp. 1–24). Springer Nature Switzerland A.G.
21. Palit, S., & Hussain, C. M., (2018). Remediation of industrial and automobile exhausts for environmental management. In: Chaudhery, M. H., (ed.), *Handbook of Environmental Materials Management* (pp. 1–17). Springer Nature Switzerland A.G.
22. Palit, S., & Hussain, C. M., (2018). Sustainable biomedical waste management. In: Chaudhery, M. H., (ed.), *Handbook of Environmental Materials Management* (pp. 1–23). Springer Nature Switzerland A.G.
23. Palit, S., & Hussain, C. M., (2018). Biopolymers, nanocomposites, and environmental protection: A far-reaching review. In: Shakeel, A., (ed.), *Bio-Based Materials for Food Packaging* (pp. 217–236). Springer Nature Singapore Pvt. Ltd., Singapore.
24. Palit, S., & Hussain, C. M., (2018). Nanocomposites in packaging: A groundbreaking review and a vision for the future. In: Shakeel, A., (ed.), *Bio-based Materials for Food Packaging* (pp. 287–303). Springer Nature Singapore Pvt. Ltd., Singapore.
25. Palit, S., (2018). Industrial vs. food enzymes: Application and future prospects. In: Mohammed, K., (ed.), *Enzymes in Food Technology: Improvements and Innovations* (pp. 319–345). Springer Nature Singapore Pvt. Ltd., Singapore.
26. Palit, S., & Hussain, C. M., (2018). Green sustainability, nanotechnology and advanced materials: A critical overview and a vision for the future. Chapter 1. In: Shakeel, A., & Chaudhery, M. H., (eds.), *Green and Sustainable Advanced Materials, Applications* (Vol. 2, pp. 1–18). Wiley Scrivener Publishing, Beverly, Massachusetts, USA.
27. Palit, S., (2018). Recent advances in corrosion science: A critical overview and a deep comprehension. In: Kharisov, B. I., (ed.), *Direct Synthesis of Metal Complexes* (pp. 379–410). Elsevier, Amsterdam, Netherlands.
28. Palit, S., (2017). Nanomaterials for industrial wastewater treatment and water purification. *Handbook of Ecomaterials* (pp. 1–41). Springer International Publishing, AG, Switzerland.

CHAPTER 9

Biocontrol by *Trichoderma* spp. as a Green Technology for the Agri-Food Industry

STEFANY ELIZABETH REZA-ESCANDÓN,[1]
CRISTÓBAL NOÉ AGUILAR,[1] RAÚL RODRÍGUEZ-HERRERA,[1]
JOSÉ D. GARCÍA-GARCÍA,[2] ANNA ILINÁ,[2]
GEORGINA MICHELENA-ÁLVAREZ,[3] and
JOSÉ LUIS MARTÍNEZ-HERNÁNDEZ[2]

[1]*Food Research Department. School of Chemistry, Autonomous University of Coahuila, Saltillo Campus, 25280 Coahuila, México*

[2]*Nanobioscience Group. Food Research Department. School of Chemistry, Autonomous University of Coahuila, Saltillo Campus, 25280 Coahuila, México, Tel:(844) 4161238, E-mail: jose-martinez@uadec.edu.mx*

[3]*Cuban Institute for Research on Sugarcane Derivatives (ICIDCA). Vía Blanca #804, Ciudad de la Habana, Zona Postal 10, Código 11000, Cuba*

ABSTRACT

Currently, agriculture depends on the use of synthetic chemicals to control phytopathogenic microorganisms that cause large losses in crops. A promising alternative to replace the use of these products is the implementation of biological control agents; such is the case of the filamentous fungi belonging to the genus *Trichoderma*. These microorganisms possess diverse mechanisms for their survival and proliferation including the mycoparasitism of phytopathogenic fungi as well as the use of complex and diverse substrates. These attributes are of great economic importance since they lead to the opportunity to be used in the production of metabolites of interest for biological control of phytopathogens under the scheme of inductive processes, besides being susceptible characteristics to the genetic improvement to potentiate its control mechanisms as it is the production of enzymes.

9.1 INTRODUCTION

The constant growth of the world population has motivated the increase of agricultural production through the intensification of agriculture. Besides, in recent times, sustainability has been prioritized, where the current agricultural challenge is not only to achieve an increase in crop yields, but also to obtain high-quality products which could be safe for health and, at the same time being eco-friendly. One of the main problems of crops is their susceptibility to attacks by various phytopathogenic microorganisms because they directly affect agricultural production. The National Research Council of the United States defines biological control as the use of natural or genetically modified organisms and that of their genes or gene products to reduce the harmful effects of plant pathogens [1]. The mechanisms of the action exerted by biological control agents belonging to the genus *Trichoderma* and the diverse range of interactions that develop in the agro-environment, show the great importance represented by the presence of enzymes, secondary metabolites and molecules involved in signaling, which isolated or together they are improving the health of the plants. The objective of this review is to describe the control mechanisms exerted by Trichoderma, recent advances in genetic improvement and inductive processes to produce enzymes that potentiate the control of phytopathogenic fungi.

9.2 PHYTOPATHOGENIC FUNGI

Fungi are eukaryotic organisms that have a cell wall and are fed by the absorption of organic molecules. They do not come from the same evolutionary trunk, most of them are multicellular except for some yeasts and produce spores for reproduction, although those known as *Mycelia sterilia* do not produce spores and most of them have a filamentous somatic structure, except plasmodioforomycetes. Encompassing microorganisms and submicroscopic entities that affect crops such as bacteria, nematodes, viruses, and viroids, fungi are the largest group, with more than 8,000 species causing diseases in plants [2]. Phytopathogenic fungi affect all crops and attack different parts of the plant causing symptoms such as chlorosis, which is the loss of green color, necrosis that refers to the death of tissue which manifests as discrete lesions, blight, rot, canker, and descending death, can also cause wilting and growth disturbances such as hyperplasia, hypertrophy, witch's broom and dwarfism [3]. One of the fungi with the greatest impact due to affecting a subsistence crop of the diet for millions of humans is *Magnaporthe grisea* (*Pyricularia grisea*) causal agent of rice blasting [4]. Other fungal diseases of great impact due to their prevalence are: *Puccinia graminis* causal agent of the rust of cereals [5], *Phytophthora infestans* that produces late blight, destructive disease of potato and tomato foliage [6], *Hemileia vastatrix* that causes coffee tree rust [7] and *Colletotrichum gloeosporioides* that causes anthracnose in tropical fruit trees [8]. There are fungi that attack a wide variety of crops, as is the case of *Rhizopus sp.* causal agent of rotting of many species of fruits and vegetables that usually develops during transport and storage [9]. On the other hand, they have been reported to some fungi that are specific in one or few hosts; such is the case of the downy mildew such as *Pseudoperonospora cubensis* that attacks cucurbits and *Peronospora destructor* to onions [10]. There are also those phytopathogens whose importance lies in the risk posed to human and animal health by the presence of mycotoxins, such as those of the genera *Aspergillus*, *Penicillium*, and *Alternaria*, fungal species that attack a great diversity of fruits, cereals, and vegetables [11].

9.3 *TRICHODERMA* IN THE CONTROL OF PHYTOPATHOGENIC FUNGI

Trichoderma is a genus of filamentous fungi of worldwide distribution abundant in the soil of diverse ecosystems and climatic zones. It includes species of great value for their ability to control foliar and soil phytopathogenic fungi that attack a wide range of plants. The control mechanisms that it exercises are: competition for space and nutrients, enzymatic antibiosis, the production of secondary metabolites and mycoparasitism. This last process includes phases such as the recognition, penetration, and death of its host. During this process, *Trichoderma* secretes enzymes that hydrolyze the cell wall of the host fungus, which has chitin as structural components, and glucans such as: β-1,3-glucans, β-1,6-glucans, and β-1,4-glucans (cellulose). These elements are embedded in a protein matrix whose intrastructural components include mannoproteins, galacto-mannoproteins, xylo-mannoproteins, glucurono-mannoproteins, and α-1,3-glucans [12], therefore, the main mechanism of antagonism against phytopathogenic fungi is the extracellular secretion of chitinase enzymes, glucanases, and proteases [13, 14]. It is also known that the presence of Trichoderma can influence the phytohormonal network of the host plant. It has been demonstrated how strains of this genus have promoted the growth of melon seedlings in the nursery, controlling the wilting by *Fusarium sp.*, where the stimulation of growth is associated with the induction of auxins [15] and the suppression of the disease to the stimulation of the defense mechanisms of the plants [16]. Table 9.1 shows examples of *Trichoderma* species that have been shown biocontrol effect against phytopathogenic fungi.

TABLE 9.1   Examples of Biocontrol of Phytopathogenic Fungi by *Trichoderma*

Trichoderma **Species**	**Phytopathogenic Fungi Target**	**References**
T. harzianum	*Alternaria alternata*	[31]
Trichoderma spp.	*Botrytis cinerea*	[32, 33]
Trichoderma spp.	*Cercospora beticola*	[34]
Trichoderma spp.	*Colletotrichum* spp.	[35–37]
Trichoderma spp.	*Fusarium* spp.	[38–41]
Trichoderma spp.	*Pythium aphanidermatum*	[42, 43]
Trichoderma spp.	*Phytophthora* spp.	[44–47]
Trichoderma spp.	*Rhizoctonia solani*	[48, 49]
Trichoderma spp.	*Rosellinia necatrix*	[50]
Trichoderma spp.	*Sclerotium rolfsii*	[51]
Trichoderma spp.	*Sclerotinia sclerotiorum*	[13, 52]
Trichoderma spp.	*Septoria tritici*	[53]
T. harzianum	*Uromyces appendiculatus*	[54]

9.4   BIOACTIVE COMPOUNDS OF *TRICHODERMA*

The control mechanisms of phytopathogens such as antibiosis, antagonism, mycoparasitism, and the induction of plant defense responses are characteristic strategies of the *Trichoderma* genus. The activity of mycoparasitism allows it to use the components of the cell wall of its prey as nutrients, whose promotion of growth results in the secretion of a diverse range of both volatile and non-volatile metabolites involved in biocontrol processes.

9.4.1 VOLATILE COMPOUNDS

Volatile organic compounds (VOCs) are the metabolites that plants and microorganisms release into the air. They are important in intra- and interspecific communication in the rhizosphere [17], being the structure infochemicals as mono- and sesquiterpenes, alcohols, ketones, lactones, esters or C8 compounds that are part of the wealth of the emissions of volatile microbial compounds. Its positive and/or negative effects on other organisms can become useful agricultural tools. It has been shown that these promote the development of plants and their defense system. Thus, in- *in vitro* tests where the plants of *Arabidopsis thaliana* exposed to the VVV of *T. virens* showed an increase in two times of the total fresh weight in comparison with the axenically cultivated seedlings. The stimulation of the branching of lateral roots was observed, thus increasing their capacity to absorb nutrients. The production of jasmonic acid, a phytohormone that occurs when plants interact with potential pathogens or with insects, was demonstrated by activating signaling cascades that increase the plant's immunity through changes in gene expression. In addition, the production of hydrogen peroxide, an oxygen reactive species that, like jasmonic acid, triggers defense responses were observed. Both help in the control of diseases, as in the case of the test carried out with plants infected with *Botrytis cinerea* that, when they were exposed to VOCs from two strains of *Trichoderma* (Tv10.4 and Tv29.8), the symptoms of induced chlorosis and the percentage of dead plants from 80% to 10 and 15% respectively were decreased [18].

Among the volatile antifungal compounds produced by *Trichoderma* species, 6-n-pentyl-2H-pyran-2-one (6-PAP) is a polyketide which is the most characteristic with a sweet coconut aroma. The analysis of the antifungal activity of 6-PAP *in vitro,* showed mycelial inhibition of *Fusarium culmorum* KF 846 in 100% for 7 days with 2.0 µg concentration of 6-PAP for each agar disc of the phytopathogenic fungus strain. This compound is involved in complex interactions against plant pathogens, as well as others, such as: 2-pentylfuran, toluene, 3-octanone, α-bergamotene, linalool isobutyrate, 2-methyl-1-propenylbenzene, β-cimene, isoamyl alcohol, 2-butanone, pentyl acetate, 3-octanol, 2-nonanone, β-bisabolene, ethyl octanoate, D-limonene, 2-heptanol, β-pinene, 2-methylbutylacetate, 1-pentanol, ethyl phenyl alcohol, α-curcumene, β-farnesene, α-cedrene, α-pinene, 1-propanol, pyridine, methyl benzoate, geranyl acetone, among others [19].

9.4.2 NON-VOLATILE COMPOUNDS

The genus *Trichoderma* produces a great variety of secondary metabolites involved in signaling processes, interaction with other organisms and biological activities that have a great capacity to control phytopathogenic fungi. As mentioned previously, the fungal cell wall is a structure mostly compound of carbohydrates (polysaccharides such as glucans, chitin, cellulose among others) and in a less proportion by proteins, lipids, ions, and pigments. *Trichoderma* species are efficient producers of hydrolytic enzymes, such as those of the chitinolytic system (chitinases and NAGase), glucanolytic (β-1,3/1,4-glucanases, α-1,3-glucanases) and proteolytic, breaking down polysaccharides and proteins which cause the loss of integrity of the cell wall which leads to the cellular collapsing [20]. Gliotoxin and glyovirin are antibiotics produced by *Trichoderma* for which their activity against phytopathogenic fungi such as *Rhizoctonia solani* has been documented. The harzianic and isoharzianic acids produced by *Trichoderma* possess antifungal activity and promote seed germination and root growth. In addition, harzianic acid is related to siderophores properties as iron-chelating agent, solubilizing it to be assimilated and used in the plant nutrient regulation mechanisms. Peptaibols, such as suzukacillins that possess antifungal activity and trichokonins that have also been shown to induce the defense system of the plant. All this, is a little part of the great diversity of soluble metabolites that represent the source of the biocontrol efficiency exerted by *Trichoderma* [21, 22].

9.5 IMPROVEMENT OF FUNGAL STRAINS AND INDUCTIVE PROCESSES

In order to have microorganisms that have characteristics that make them suitable as good biocontrol agents and producers of enzymes, alternatives have been sought to overcome the marks defined by nature. This is the reason why researchers have focused in carry out studies in which microorganisms are mutated. Such is the case of the improvement of *T. harzianum* CECT 2413 in which a conidial suspension was mutagenized with N-methyl-N'-N-nitrosoguanidine. The result was an increase in enzymatic chitinase activity above the parental strain and double increases in enzymatic activity β-1,3 and β-1,6-glucanase. In addition, a threefold increase in the production of extracellular proteins was observed when using culture media with inducing agents, improving sporulation and control against *Rhizoctoniasolani* in a shorter time, together with the protection of grapes against *Botrytis cinerea* in- *in vitro* trials [23]. As a response to the needing of having fungal strains of improved efficiencies, in addition to their implementation to generate the reduction of production costs, work has been done to obtain strains such as the recombinant *T. reesei* TRB1, developed from *T. reesei* RUT-C30 by mutagenesis based on T-DNA. This recombinant strain showed a significant improvement in extracellular β-glucosidase activity with a 17-fold hyperproduction, as well as high production of cellulase in a shorter time, making it suitable for being used in industrial production processes [24]. Other examples of the transformation of genes as a biocontrol strategy that has shown favorable results, lies in the study in which the transfer of the *chiV* gene to the fungus *T. harzianum* takes place. The genetically modified microorganism could control the viability of *Rhizoctoniasolani*, increasing its inhibition from 82.4% to 98.5% [25].

The SOD (superoxide dismutase) gene was transferred to *T. harzianum*. It was observed that the modified strain after being subjected to stress conditions due to temperature and salinity increasing (40°C and NaCl 2 mol/L at 27°C for 5–10 days) was able to inhibit the growth of *S. sclerotiorum* in 83.9% and 60.1% in the respective tests, unlike the parental strain which did not survive in the face of said adverse conditions [26].

With the modification of *T. reesei* by the heterogonous expression of a lipase gene from *Talaromyces thermophilus*, the strain could produce lipases with high alkaline tolerance and thermostability with enzymatic activity of 241 IU/ml, retaining more than 70% activity at pH 11 or exposed to 70°C for 1 h [27].

On the other hand, it is known that the production of enzymes is sensitive to the composition of the culture medium and the presence of the inducing compounds. Thus, the response of a strain of *T. harzianum* ETS 323 was evaluated in the presence of inactivated mycelium of phytopathogenic fungi added to the culture medium of the strain. They compared the production of β-1,3-glucanase enzymes, β-1,6-glucanases, chitinases, proteases, and xylanases obtained in culture media which contained inactivated mycelium of *Botrytis cinerea* or *T. harzianum*. Outcomes showed greater activity in the culture medium containing inactivated mycelium of *B. cinerea* as the sole carbon source, being in the same medium in which the induction of

L-amino acid oxidase (LAAO) and 2 endochitinases was detected which were not found in the other fermentation trials [28].

In the studies performed with a strain of *T. atroviride* native and transgenic with the endochitinase gene (ThEn-42) activated by the cellulase promoter cbh1 of *T. reesei* for overexpression of ThEn-42, 8 different culture media were evaluated. It was found that the carbon source played an important role in the expression of the enzymes of interest for biological control. The culture media characterized by having as a source of carbon chitin and cellulose, respectively, were those that allowed obtaining extracts with very good antifungal activity against the germination of *Penicillium digitatum*. In addition, it was demonstrated that trace elements played a determining role in protein profiles, observing trends in the production of extracts with greater antifungal activities in culture media containing elements such as Cu and Mo [29].

9.6 VALORIZATION OF SUBSTRATES FOR THE OBTAINING OF BIOACTIVE MOLECULES

The processing of raw materials from agro-industry generates a large quantity and a diverse range of by-products. *Trichoderma* has demonstrated the ability to take advantage of plant waste derived from the processing of fruits and vegetables in submerged culture [30], this type of waste, as well as others, can be used in biotechnological processes, which have been shown to be an area of opportunity.

The increase in the production of biopesticides, in response to their growing demand in recent years, has led to the search for suitable strategies for their production. The use of agroindustrial waste as substrates, the reduction in energy consumption, the low generation of wastewater, and the ability to obtain products with high stability are qualities that solid fermentation technology offers [67]. The fermentation in a solid medium is defined as a microbial process that usually occurs in solid materials in the absence of free water. Its advantages compared to those provided by fermentation in a liquid medium, represents an economic and effective alternative for the generation of biomolecules of interest in biological control with a high added value. The production of enzymes for biocontrol from various species of *Trichoderma* by fermentation in solid medium taking advantage of agroindustrial residues as well as various inductive substrates is shown in Table 9.2.

TABLE 9.2 Enzymes Production for Biocontrol by Different Species of *Trichoderma* Trough Fermentation in Solid State with Agroindustry Waste and Inducing Substrates

Strain	Substrate	Reactor	Chitinase (U/gds)		B-1,3-Glucanase (U/gds)		Cellulase (U/gds)		Protease (U/gds)	Lipase (U/gds)	References
			Endo	Exo	Endo	Exo	Endo	Exo			
T. longibranchiatum	Bagasse of sugarcane, wheat bran, chitin, potato flour, and olive oil	Glass columns		30.74				9.29		11.56	[55]
T. yunnanense							38.32				
T. viride WEBL0703	Grape pomace and wine lees	Erlenmeyer flasks		47.8		8.32					[56]
T. harzianum TUBF691	Wheat bran with colloidal chitin	Erlenmeyer flask		10.7							[57]
T. longibrachiatum G26	Corncobs	Erlenmeyer flask						136.2			[58]
Trichoderma sp.	Crude bagasse of walnut palm	Erlenmeyer flasks						6.9			[59]
	Degreased bagasse of walnut palm							16.1			
	Vegetables waste							50.1			
T. reesei	Corn residues	Jars			20.32			1.84			[60]
	Sunflower residue				10.56			1.6			
T. atroviride T42	Mushroom waste	Erlenmeyer flasks					0.76				[61]
T. konigii TK1							0.3				
T. harzianum T 10							0.12				

TABLE 9.2 *(Continued)*

Strain	Substrate	Reactor	Chitinase (U/gds)		B-1,3-Glucanase (U/gds)		Cellulase (U/gds)		Protease (U/gds)	Lipase (U/gds)	References
			Endo	Exo	Endo	Exo	Endo	Exo			
T. reesei DSMZ 970	Potato peals	Erlenmeyer flasks						41.8			[62]
T. reeseiRut-30	Citrus residues with wheat bran	Erlenmeyer flasks					55	25	23		[63]
T. asperellumT2-10	Corncob with *C. gloeosporioides* biomass	Polyethylen Bags	5.11					2.58			[55]
	Corncob with *P. capsici* biomass		3.6					7.82			
Trichoderma sp. T-II	Mango peals with salts	Erlenmeyer flasks						7.5			[64]
T. harzianum FCLA 501	Castor bagasse	Erlenmeyer flasks								2.16	[65]
	Sugarcane bagasse									0.20	
	Castor and yucca bagasse									2.70	
	Castor bagasse and corn residues									2.70	
	Castor and sugarcane bagasse									4.04	

9.7 CONCLUSION

Trichoderma species are found in the ecosystem, naturally suppressing phytopathogenic fungi. The genus produces degrading enzymes of the fungal cell wall which leads to mycoparasitism, promoting the reduction of the densities of phytopathogenic fungi in the agro-environment. The use of agroindustrial waste using technologies such as fermentation in solid media allows the obtaining of said biomolecules in a sustainable manner, a process in which the selection of suitable substrates and convenient microorganisms are the key to obtain extracts with greater antifungal activities. The implementation of genetically improved strains of *Trichoderma* and the use of inducing agents in the culture medium are strategies that allow maximizing the production of enzymes for biocontrol, which are well suited to be used in production processes.

ACKNOWLEDGMENTS

Authors thank CONACYT for the scholarship awarded to Stefany Elizabeth Reza Escandón during the completion of her graduate studies.

KEYWORDS

- **agroindustrial waste**
- **fermentation**
- **mycoparasitism**
- *Penicillium digitatum*
- **phytopathogenic fungi**
- *Trichoderma*

REFERENCES

1. Babbal, A., & Khasa, Y. P., (2017). Microbes as biocontrol agents. In: Kumar, V., Kumar, M., Sharma, S., & Prasad, R., (eds.), *Probiotics and Plant Health*

(pp. 507–552, 600). Springer Singapore. Singapore. https://doi.org/10.1007/978-981-10-3473-2_24 (accessed on 22 February 2020).

2. Agrios, G. N., (2006). Plant diseases caused by fungi. In: Grupo, N., (ed.), *Fitopatología* (pp. 273–530, 838). Editorial Limusa S. A. De C.V., 1995, México.

3. Arauz, L. F., (1998). Phytopathogenic Fungi. In: Carazo, G., Murillo, M., Fernández, G., Brenes, S., Ramírez, L., & Araya, C., (eds.), *Phytopathology: An Agroecological Approach. Editorial Costa Rica University* (pp. 105–136, 471). Costa Rica.

4. Tosa, Y., & Chuma, I., (2014). Classification and parasitic specialization of blast fungi. *Journal of General Plant Pathology, 80*, 202–209. https://doi.org/10.1007/s10327-014-0513-7 (accessed on 22 February 2020).

5. Szabo, L. J., Cuomo, C. A., & Park, R. F., (2014). *Puccinia graminis*. In: Dean, R., Lichens-Park, A., & Kole, C., (eds.), *Genomics of Plant-Associated Fungi: Monocot Pathogens* (pp. 177–196, 201). Springer-Verlag Berlin Heidelberg. Alemania. https://doi.org/10.1007/978-3-662-44053-7_8 (accessed on 22 February 2020).

6. Judelson, H. S., (2014). *Phytophthora infestans*. In: Dean, R., Lichens-Park, A., & Kole, C., (eds.), *Genomics of Plant-Associated Fungi and Oomycetes: Dicot Pathogens* (pp. 175–208, 239). Springer, Berlin, Heidelberg. Alemania. https://doi.org/10.1007/978-3-662-44056-8_9 (accessed on 22 February 2020).

7. Capucho, A. S., Zambolim, E. M., Freitas, R. L., et al., (2012). Identification of race XXXIII of *Hemileia vastatrix* on *Coffea arabica* catimor derivatives in Brazil. *Australasian Plant Disease Notes, 7*, 189–191. https://doi.org/10.1007/s13314-012-0081-7 (accessed on 22 February 2020).

8. Kamle, M., & Kumar, P., (2016). *Colletotrichum gloeosporioides*: Pathogen of anthracnose disease in mango (*Mangifera indica* L.). In: Kumar, P., Gupta, V., Tiwari, A., & Kamle, M., (eds.), *Current Trends in Plant Disease Diagnostics and Management Practices, Fungal Biology* (pp. 207–219, 469). Springer, Cham. https://doi.org/10.1007/978-3-319-27312-9_9 (accessed on 22 February 2020).

9. Pitt, J. I., & Hocking, A. D., (2009). Fresh and perishable foods. In: *Fungi and Food Spoilage* (pp. 383–400, 519). Springer, Boston, MA. USA. https://doi.org/10.1007/978-0-387-92207-2_11 (accessed on 22 February 2020).

10. Horst, R. K., (2013). Downy mildews. In: *Department of Plant Pathology and Plant-Microbe Biology, Westcott's Plant Disease Handbook* (pp. 181–186, 826). Cornell University, Springer Science+Business Media Dordrecht. https://doi.org/10.1007/978-94-007-2141-8_27 (accessed on 22 February 2020).

11. Sanzani, S. M., Reverberi, M., & Geisen, R., (2016). Mycotoxins in harvested fruits and vegetables: Insights in producing fungi, biological role, conducive conditions, and tools to manage postharvest contamination. *Postharvest Biology and Technology, 122*, 95–105. https://doi.org/10.1016/j.postharvbio.2016.07.003 (accessed on 22 February 2020).

12. Feofilova, E. P., (2010). The fungal cell wall: Modern concepts of its composition and biological function. *Microbiology, 79*, 711–720. https://doi.org/10.1134/S00262617 10060019 (accessed on 22 February 2020).

13. Geraldine, A. M., Lopes, F. A. C., Carvalho, D. D. C., et al., (2013). Cell wall-degrading enzymes and parasitism of sclerotia are key factors on field biocontrol of white mold by *Trichoderma* spp. *Biological Control, 67*, 308–316. https://doi.org/10.1016/j.biocontrol.2013.09.013 (accessed on 22 February 2020).

14. Qualhato, T. F., Lopes, F. A. C., Steindorff, A. S., et al., (2013). Mycoparasitism studies of *Trichoderma* species against three phytopathogenic fungi: Evaluation of antagonism and hydrolytic enzyme production. *Biotechnology Letters, 35,* 1461–1468. https://doi. org/10.1007/s10529-013-1225-3 (accessed on 22 February 2020).

15. Martínez-Medina, A., Del Mar Alguacil, M., Pascual, J. A., et al., (2014). Phytohormone profiles induced by *Trichoderma* isolates correspond with their biocontrol and plant growth-promoting activity on melon plants. *Journal of Chemical Ecology, 40,* 804–815. https://doi.org/10.1007/s10886-014-0478-1 (accessed on 22 February 2020).

16. Pandey, V., Shukla, A., & Kumar, J., (2016). Physiological and molecular signaling involved in disease management through *Trichoderma*: An effective biocontrol paradigm. In: Kumar, P., Gupta, V., Tiwari, A., & Kamle, M., (eds.), *Current Trends in Plant Disease Diagnostics and Management Practices, Fungal Biology,* (pp. 317–346, 469). Springer, Cham. https://doi.org/10.1007/978-3-319-27312-9_14 (accessed on 22 February 2020).

17. Van Dam, N. M., Weinhold, A., & Garbeva, P., (2016). Calling in the dark: The role of volatiles for communication in the rhizosphere. In: Blande, J., & Glinwood, R., (eds.), *Deciphering Chemical Language of Plant Communication* (pp. 175–210, 326). Signaling and Communication in Plants. Springer International Publishing Switzerland. https://doi.org/10.1007/978-3-319-33498-1_8 (accessed on 22 February 2020).

18. Contreras-Cornejo, H. A., Macías-Rodríguez, L., Herrera-Estrella, A., et al., (2014). The 4-phosphopantetheinyl transferase of *Trichoderma virens* plays a role in plant protection against *Botrytis cinerea* through volatile organic compound emission. *Plant and Soil, 379,* 261–274. https://doi.org/10.1007/s11104-014-2069-x (accessed on 22 February 2020).

19. Jeleń, H., Błaszczyk, L., Chełkowski, J., et al., (2014). Formation of 6-n-pentyl-2H-pyran-2-one (6-PAP) and other volatiles by different *Trichoderma* species. *Mycological Progress, 13,* 589–600. https://doi.org/10.1007/s11557-013-0942-2 (accessed on 22 February 2020).

20. Viterbo, A., Ramot, O., Chernin, L., et al., (2002). Significance of lytic enzymes from *Trichoderma* spp. in the biocontrol of fungal plant pathogens. *Antonie van Leeuwenhoek, 81,* 549–556. https://doi.org/10.1023/A:1020553421740 (accessed on 22 February 2020).

21. Keswani, C., Mishra, S., Sarma, B. K., et al., (2014). Unraveling the efficient applications of secondary metabolites of various *Trichoderma* spp. *Applied Microbiology and Biotechnology, 98,* 533–544. https://doi.org/10.1007/s00253-013-5344-5 (accessed on 22 February 2020).

22. Patil, A. S., Patil, S. R., & Paikrao, H. M., (2016). *Trichoderma* secondary metabolites: Their biochemistry and possible role in disease management. In: Choudhary, D., & Varma, A., (eds.), *Microbial-Mediated Induced Systemic Resistance in Plants* (pp. 69–102, 226). Springer Science+Business Media Singapore. Singapore. https://doi. org/10.1007/978-981-10-0388-2_6 (accessed on 22 February 2020).

23. Rey, M., Delgado-Jarana, J., & Benítez, T., (2001). Improved antifungal activity of a mutant of *Trichoderma harzianum* CECT 2413 which produces more extracellular proteins. *Applied Microbiology and Biotechnology, 55,* 604–608. https://doi.org/ 10.1007/s002530000551 (accessed on 22 February 2020).

24. Li, C., Lin, F., Li, Y., et al., (2016). A β-glucosidase hyper-production *Trichoderma reesei* mutant reveals a potential role of cel3D in cellulase production. *Microbial*

Cell Factories, 15, 151. https://doi.org/10.1186/s12934-016-0550-3 (accessed on 22 February 2020).

25. Yang, L., Yang, Q., Sun, K., et al., (2011). *Agrobacterium tumefaciens* mediated transformation of ChiV gene to *Trichoderma harzianum. Applied Biochemistry and Biotechnology, 163*, 937–945. https://doi.org/10.1007/s12010-010-9097-7 (accessed on 22 February 2020).

26. Yang, L., Yang, Q., Sun, K., et al., (2010). *Agrobacterium tumefaciens*-mediated transformation of SOD gene to *Trichoderma harzianum. World Journal of Microbiology and Biotechnology, 26*, 353–358. https://doi.org/10.1007/s11274-009-0182-4 (accessed on 22 February 2020).

27. Zhang, X., Li, X., & Xia, L., (2015). Heterologous expression of an alkali and thermotolerant lipase from *Talaromyces thermophilus* in *Trichoderma reesei. Applied Biochemistry and Biotechnology, 176*, 1722–1735. https://doi.org/10.1007/s12010-015-1673-4 (accessed on 22 February 2020).

28. Yang, H. H., Yang, S. L., Peng, K. C., et al., (2009). Induced proteome of *Trichoderma harzianum* by *Botrytis cinerea. Mycological Research, 113*, 924–932. https://doi.org/10.1016/j.mycres.2009.04.004 (accessed on 22 February 2020).

29. Deng, S., Lorito, M., Penttilä, M., et al., (2007). Overexpression of an endochitinase gene (*ThEn-42*) in *Trichoderma atroviride* for increased production of antifungal enzymes and enhanced antagonist action against pathogenic fungi. *Applied Biochemistry and Biotechnology, 142*, 81–94. https://doi.org/10.1007/s12010-007-0012-9 (accessed on 22 February 2020).

30. Ma'tat'a, M., Cibulová, A., Varečka, L., et al., (2016). Plant waste residues as inducers of extracellular proteases for a deuteromycete fungus *Trichoderma atroviride. Chemical Papers, 70*, 1039–1048. https://doi.org/10.1515/chempap-2016-0040 (accessed on 22 February 2020).

31. Sempere, F., & Santamarina, M. P., (2007). *In vitro* biocontrol analysis of *Alternaria alternata* (Fr.) Keissler under different environmental conditions. *Mycopathologia, 163*, 183–190. https://doi.org/10.1007/s11046-007-0101-x (accessed on 22 February 2020).

32. Dik, A., & Elad, Y., (1999). Comparison of antagonists of *Botrytis cinerea* in greenhouse-grown cucumber and tomato under different climatic conditions. *European Journal of Plant Pathology, 105*, 123–137. https://doi.org/10.1023/A:1008778213278 (accessed on 22 February 2020).

33. Tronsmo, A., & Dennis, C., (1977). The use of *Trichoderma* species to control strawberry fruit rots. *Netherlands Journal of Plant Pathology, 83*, 449–455. https://doi.org/10.1007/BF03041462 (accessed on 22 February 2020).

34. Galletti, S., Burzi, P. L., Cerato, C., et al., (2008). *Trichoderma* as a potential biocontrol agent for *Cercospora* leaf spot of sugar beet. *BioControl, 53*, 917–930. https://doi.org/10.1007/s10526-007-9113-1 (accessed on 22 February 2020).

35. Freeman, S., Minz, D., Kolesnik, I., et al., (2004). *Trichoderma* biocontrol of *Colletotrichum acutatum* and *Botrytis cinerea* and survival in strawberry. *European Journal of Plant Pathology, 110*, 361–370. https://doi.org/10.1023/B:EJPP.0000021057.93305.d9 (accessed on 22 February 2020).

36. De Los Santos-Villalobos, S., Guzmán-Ortiz, D. A., Gómez-Lim, M. A., et al., (2013). Potential use of *Trichoderma asperellum* (Samuels, Liechfeldt et Nirenberg)

T8a as a biological control agent against anthracnose in mango (*Mangifera indica* L.). *Biological Control, 64*, 37–44. https://doi.org/10.1016/j.biocontrol.2012.10.006 (accessed on 22 February 2020).

37. Landero, V. N., Nieto, A. D., Téliz, O. D., et al., (2015). Biological control of anthracnose by postharvest application of *Trichoderma* spp. on maradol papaya fruit. *Biological Control, 91*, 88–93. https://doi.org/10.1016/j.biocontrol.2015.08.002 (accessed on 22 February 2020).

38. Marzano, M., Gallo, A., & Altomare, C., (2013). Improvement of biocontrol efficacy of *Trichoderma harzianum* vs. *Fusarium oxysporum* f. sp. *Lycopersici* through UV-induced tolerance to fusaric acid. *Biological Control, 67*, 397–408. https://doi.org/10.1016/j.biocontrol.2013.09.008 (accessed on 22 February 2020).

39. Ferrigo, D., Raiola, A., Rasera, R., & Causin, R., (2014). *Trichoderma harzianum* seed treatment controls *Fusarium verticillioides* colonization and fumonisin contamination in maize under field conditions. *Crop Protection, 65*, 51–56. https://doi.org/10.1016/j.cropro.2014.06.018 (accessed on 22 February 2020).

40. De Souza, J. T., Trocoli, R. O., & Monteiro, F. P., (2016). Plants from the Caatinga biome harbor endophytic *Trichoderma* species active in the biocontrol of pineapple fusariosis. *Biological Control, 94*, 25–32. https://doi.org/10.1016/j.biocontrol.2015.12.005 (accessed on 22 February 2020).

41. Toghueo, R. K., Eke, P., Zabalgogeazcoa, I., et al., (2016). Biocontrol and growth enhancement potential of two endophytic *Trichoderma* spp. From *Terminalia catappa* against the causative agent of common bean root rot (*Fusarium solani*). *Biological Control, 96*, 8–20. https://doi.org/10.1016/j.biocontrol.2016.01.008 (accessed on 22 February 2020).

42. Kamala, T., & Indira, S., (2011). Evaluation of indigenous *Trichoderma* isolates from Manipur as biocontrol agent against *Pythium aphanidermatum* on common beans. *Three Biotech., 1*, 217–225. https://doi.org/10.1007/s13205-011-0027-3 (accessed on 22 February 2020).

43. Sain, S. K., & Pandey, A. K., (2016). Evaluation of Some *Trichoderma harzianum* isolates for the management of soilborne diseases of brinjal and okra. *Proceedings of the National Academy of Sciences, India Section B: Biological Sciences, 1*, 1–10. https://doi.org/10.1007/s40011-016-0824-x (accessed on 22 February 2020).

44. Mbarga, J. B., Begoude, B. A. D., Ambang, Z., et al., (2014). A new oil-based formulation of *Trichoderma asperellum* for the biological control of cacao black pod disease caused by *Phytophthora megakarya*. *Biological Control, 77*, 15–22. https://doi.org/10.1016/j.biocontrol.2014.06.004 (accessed on 22 February 2020).

45. Widmer, T. L., (2014). Screening *Trichoderma* species for biological control activity against *Phytophthora ramorum* in soil. *Biological Control, 79*, 43–48. https://doi.org/10.1016/j.biocontrol.2014.08.003 (accessed on 22 February 2020).

46. Sriwati, R., Melnick, R. L., Muarif, R., et al., (2015). *Trichoderma* from Aceh Sumatra reduce *Phytophthora* lesions on pods and cacao seedlings. *Biological Control, 89*, 33–41. https://doi.org/10.1016/j.biocontrol.2015.04.018 (accessed on 22 February 2020).

47. Jiang, H., Zhang, L., Zhang, J., et al., (2016). Antagonistic interaction between *Trichoderma asperellum* and *Phytophthora capsici in vitro*. *Journal of Zhejiang University-Science B, 17*, 271–281. https://doi.org/10.1631/jzus.B1500243 (accessed on 22 February 2020).

48. Anees, M., Tronsmo, A., Edel-Hermann, V., et al., (2010). Characterization of field isolates of *Trichoderma* antagonistic against *Rhizoctonia solani*. *Fungal Biology, 114*, 691–701. https://doi.org/10.1016/j.funbio.2010.05.007 (accessed on 22 February 2020).

49. Kotasthane, A., Agrawal, T., Kushwah, R., et al., (2015). In-vitro antagonism of *Trichoderma* spp. against *Sclerotium rolfsii* and *Rhizoctonia solani* and their response towards growth of cucumber, bottle gourd, and bitter gourd. *European Journal of Plant Pathology, 141*, 523–543. https://doi.org/10.1007/s10658-014-0560-0 (accessed on 22 February 2020).

50. Freeman, S., Sztejnberg, A., & Chet, I., (1986). Evaluation of *Trichoderma* as a biocontrol agent for *Rosellinia necatrix*. *Plant and Soil, 94*, 163–170. https://doi.org/10.1007/BF02374340 (accessed on 22 February 2020).

51. John, N. S., Anjanadevi, I. P., Nath, V. S., et al., (2015). Characterization of *Trichoderma* isolates against *Sclerotium rolfsii*, the collar rot pathogen of *Amorphophallus*: A polyphasic approach. *Biological Control, 90*, 164–172. https://doi.org/10.1016/j.biocontrol.2015.07.001 (accessed on 22 February 2020).

52. Troian, R. F., Steindorff, A. S., Ramada, M. H. S., et al., (2014). Mycoparasitism studies of *Trichoderma harzianum* against *Sclerotinia sclerotiorum*: evaluation of antagonism and expression of cell wall-degrading enzymes genes. *Biotechnology Letters, 36*, 2095–2101. https://doi.org/10.1007/s10529-014-1583-5 (accessed on 22 February 2020).

53. Perelló, A. E., Moreno, M. V., Mónaco, C., et al., (2009). Biological control of *Septoria tritici* blotch on wheat by *Trichoderma* spp. under field conditions in Argentina. *BioControl, 54*, 113–122. https://doi.org/10.1007/s10526-008-9159-8 (accessed on 22 February 2020).

54. Burmeister, L., & Hau, B., (2009). Control of the bean rust fungus *Uromyces appendiculatus* by means of *Trichoderma harzianum*: Leaf disc assays on the antibiotic effect of spore suspensions and culture filtrates. *BioControl., 54*, 575–585. https://doi.org/10.1007/s10526-008-9202-9 (accessed on 22 February 2020).

55. De la Cruz-Quiroz, R., Robledo-Padilla, F., Aguilar, C. N., et al., (2017). Forced aeration influence on the production of spores by *Trichoderma* strains. *Waste and Biomass Valorization, 8*, 2263–2270. https://doi.org/10.1007/s12649-017-0045-4 (accessed on 22 February 2020).

56. Bai, Z., Jin, B., Li, Y., et al., (2008). Utilization of winery wastes for *Trichoderma viride* biocontrol agent production by solid state fermentation. *Journal of Environmental Sciences, 20*, 353–358. https://doi.org/10.1016/S1001-0742(08)60055-8 (accessed on 22 February 2020).

57. Sandhya, C., Binod, P., Nampoothiri, K. M., et al., (2005). Microbial synthesis of chitinase in solid cultures and its potential as a biocontrol agent against phytopathogenic fungus *Colletotrichum gloeosporioides*. *Applied Biochemistry and Biotechnology, 127*, 1–15. https://doi.org/10.1385/ABAB:127:1:001 (accessed on 22 February 2020).

58. Xie, L., Zhao, J., Wu, J., et al., (2015). Efficient hydrolysis of corncob residue through cellulolytic enzymes from *Trichoderma* strain G26 and l-lactic acid preparation with the hydrolysate. *Bioresource Technology, 193*, 331–336. https://doi.org/10.1016/j.biortech.2015.06.101 (accessed on 22 February 2020).

59. Lah, T. N. T., Norulaini, N. A. N., Shahadat, M., et al., (2016). Utilization of industrial waste for the production of cellulase by the cultivation of *Trichoderma* via solid state

fermentation. *Environmental Processes, 3*, 803–814. https://doi.org/10.1007/s40710-016-0185-8 (accessed on 22 February 2020).

60. Safari, S. A. A., Ghanbari, M., & Janjan, A., (2009). Improvement of digestibility of sunflower and corn residues by some saprophytic fungi. *Journal of Material Cycles and Waste Management, 11*, 293–298. https://doi.org/10.1007/s10163-009-0245-5 (accessed on 22 February 2020).

61. Grujić, M., Dojnov, B., Potočnik, I., et al., (2015). Spent mushroom compost as substrate for the production of industrially important hydrolytic enzymes by fungi *Trichoderma* spp. and *Aspergillus niger* in solid state fermentation. *International Biodeterioration and Biodegradation, 104*, 290–298. https://doi.org/10.1016/j.ibiod.2015.04.029 (accessed on 22 February 2020).

62. Ben, T. I., Bennour, H., Fickers, P., et al., (2017). Valorization of potato peels residues on cellulase production using a mixed culture of *Aspergillus niger* ATCC 16404 and *Trichoderma reesei* DSMZ 970. *Waste and Biomass Valorization, 8*, 183–192. https://doi.org/10.1007/s12649-016-9558-5 (accessed on 22 February 2020).

63. Oberoi, H. S., Babbar, N., Dhaliwal, S. S., et al., (2012). Enhanced oil recovery by pre-treatment of mustard seeds using crude enzyme extract obtained from mixed-culture solid-state fermentation of kinnow (*Citrus reticulata*) waste and wheat bran. *Food and Bioprocess Technology, 5*, 759–767. https://doi.org/10.1007/s11947-010-0380-y (accessed on 22 February 2020).

64. Buenrostro-Figueroa, J., De La Garza-Toledo, H., Ibarra-Junquera, V., et al., (2010). Juice extraction from mango pulp using an enzymatic complex of *Trichoderma* sp. produced by solid-state fermentation. *Food Science and Biotechnology, 19*, 1387–1390. https://doi.org/10.1007/s10068-010-0197-5 (accessed on 22 February 2020).

65. Coradi, G. V., Da Visitação, V. L., De Lima, E. A., et al., (2013). Comparing submerged and solid-state fermentation of agro-industrial residues for the production and characterization of lipase by *Trichoderma harzianum. Annals of Microbiology, 63*, 533–540. https://doi.org/10.1007/s13213-012-0500-1 (accessed on 22 February 2020).

66. De la Cruz-Quiroz, R, Roussos, S, Hernández-Castillo, D., et al., (2015). Challenges and opportunities of the bio-pesticides production by solid-state fermentation: Filamentous fungi as a model. *Critical Reviews in Biotechnology, 35*, 326–333. http://dx.doi.org/10.3109/07388551.2013.857292 (accessed on 22 February 2020).

67. De la Cruz-Quiroz, R, Roussos, S, Hernandez-Castillo, D., et al., (2017). Solid-state fermentation in a bag bioreactor: Effect of corn cob mixed with phytopathogen biomass on spore and cellulase production by *Trichoderma asperellum*. In: Jozala, A. F., (ed.), *Fermentation Processes* (pp. 43–56, 310). InTech. http://dx.doi.org/10.5772/64643 (accessed on 22 February 2020).

CHAPTER 10

Phenomena of Charge Quantization, Interference, and Annihilation in Chemical Reactions of Mesoparticles

V. I. KODOLOV,[1,2] V. V. KODOLOVA-CHUKHONTSEVA,[1,3]
N. S. TEREBOVA,[1,3] I. N. SHABANOVA,[1,3] YU. V. PERSHIN,[1]
R. V. MUSTAKIMOV,[1,2] and A. YU. BONDAR[2]

[1]*Basic Research-High Educational Center of Chemical Physics and Mesoscopics, Izhevsk, Russia*

[2]*M.T. Kalashnikov Izhevsk State Technical University, Izhevsk, Russia*

[3]*Udmurt Federal Research Center, Ural Division, Russian Academy of Sciences, Izhevsk, Russia*

ABSTRACT

In this chapter such main phenomena as charge quantization, interference and annihilation which are characteristic for mesoscopics at the realization of chemical reactions without change and with change of oxidation states of elements participating are considered. It's noted that the interference leads to the chemical bonds formation and takes place in the reactions without changes of the oxidation states. In contrast with preceding the annihilation is occurred at the Red Ox reactions when the changes of oxidation states are realized. This phenomenon leads to the electron shifts and also to the magnetic characteristics increasing as well as to the creation of interference with the chemical bonds appearance. The investigations are carried out on the examples of analysis for processes of copper or nickel-carbon mesocomposites modification by the compounds containing p, d elements. In the middle of such substances as polyethylene polyamine, ammonium iodide, ammonium polyphosphate (APP), silica (SiO_2), aluminum oxide, iron oxide, nickel oxide and copper oxide are used. In the case, when polyethylene polyamine and ammonium iodide are applied, the connection reactions takes place. At the interactions of polyethylene polyamine with mesoparticles the C=N bond formation is explained by the interference of negative charges quants. When the mesoparticles modification reactions with the using APP, SiO_2, metal oxides are carried out, the red ox processes are realized. The hypothesis concerning to the passing of two phenomena (annihilation and interference) at red ox processes is proposed.

10.1   INTRODUCTION

New scientific trend-chemical mesoscopics on notion reflects the mesoparticles (mesosystems) reactivity and the kinetics of mesoparticles reactions. In this trend, the following peculliarities of mesoscopics phenomena are considered such as charges quantization, interference, and annihilation. It's possible there is not only quantization of negative charges but also positive charge quantization. If there is many information about electron wave nature, then the information concerning to positron wave nature is very small. For example, it's known that the positron can be quantizated near nucleus [2]. Many questions arise about the nature of positive charge quants and formation of them. However, on the analogy with planet macrosystem it's possible to suppose the creation of positive charge quants flow in the space near atom nucleus. According to [3] the formation of chemical bonds is explained by the interference of electron waves. Therefore, the energy of bond is written as the sum of following energies: interference energy, energy of partition-penetration, energy of surroundings (energy of neighbor influence). The notion of partition-penetration energy can be presented as the mutual penetration of electron waves accompanied with partition on quants. The third constituent part of energy sum is determined by pair interaction of atoms "radiated" contrary electron waves. This explanation of chemical bond formation is correspondent to the classic idea about the formation of high electron density between atoms (chemical bond).

   If chemical reactions are, realize without the changes of atoms oxidation states (OSs) then the negative charges quants quantization and the interference are carried out. However, the most reactions flow with the changes of elements OSs and then according to known schemes of reduction-oxidation processes, it's necessary to take into consideration of positive charge quants. At the interaction of positive charge quants with the negative charge quants the annihilation phenomenon with the electromagnetic radiation or/and the direct electromagnetic field is possible. Also, the interaction of positive charges with the formation of "dark hole" must not exclude (Figure 10.1). In this case, the explosion with diffusion of many most quantity of energy into surroundings is possible.

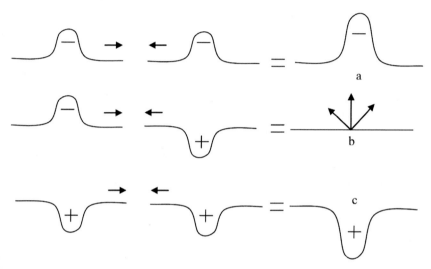

FIGURE 10.1 (a) The scheme of charge quants interaction with appearance of interference, (b) annihilation and (c) the formation of "dark hole."

The phenomena of charge quantization, interference, and annihilation are considered on the examples of metal/carbon nanocomposites (NCs) (mesoparticles) interactions with reagents containing p, d elements. At these investigations, the basic method of researches is x-ray photoelectron spectroscopy.

10.2 DESCRIPTION OF COPPER OR NICKEL CONTAINING MESOPARTICLES USED AS REAGENTS IN REACTIONS WITHOUT CHANGE AND WITH CHANGE OF ELEMENTS OXIDATION STATES (OSS)

In our investigations, it's established that the metal clusters within metal/carbon mesoparticles have form near to spheric and are found in carbon shells. The middle size of metal cluster corresponds to 20–30 nm for copper/carbon nanocomposite (Cu/C NC) and 19–20 nm for nickel/carbon nanocomposite (Ni/C NC).

The structures of carbon shells are defined by means of the complex of methods including x-ray photoelectron spectroscopy, transition electron microscopy with high permission, electron microdiffraction (Figures 10.2 and 10.3).

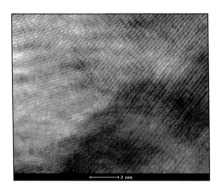 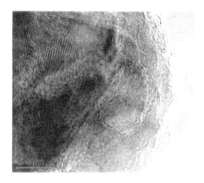

FIGURE 10.2   Microphotographs of TEM for Cu/C NC (a) and Ni/C NC (b).

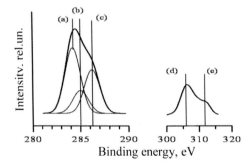 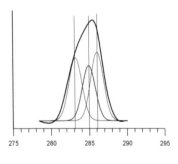

FIGURE 10.3   C1s spectra for Cu/C NC (a) and Ni/C NC (b).

According to Figure 10.2, the carbon structure shells for clusters of copper and nickel have near forms and are presented as carbon fibers. This fact is confirmed by electron microdiffraction results. In correspondence with C1s spectra (Figure 10.3) the carbon fibers contain the carbine and polyacetylene fragments: peak at 285 eV (C-H bond) and addition to peak at 281–282 eV (carbine).

It's possible to suppose that the unpaired electrons are found on the joints between carbine and polyacetylene fragments. The stability of formed systems is provided for metal coordination with double bonds, mainly with carbine bonds.

From the comparison of C1s spectra for copper and nickel/carbon NCs the peak C-H bond excess on 30% for Ni/C NC on the relation correspondent peak for Cu/C NC. This result can be explained by the decreasing of carbine fragments in carbon shell of Ni/C NC that is corresponded with incomplete reduction of nickel oxide at the obtaining of this NC (Table 10.1).

TABLE 10.1 Composition of Metal Containing Phases in Metal/Carbon Nanocomposites

Phase	Cu/C Nanocomposite	Ni/C Nanocomposite
CuO	1.17%	–
Cu$_2$O	5.19%	–
Cu	93.64%	–
NiO	–	32.15%
Ni	–	67.85%

Therefore the quantity of joints between carbine and polyacetylene fragments for Ni/C NC is bigger than for Cu/C NC, and also unpaired electron excess for Ni/C NC more that is confirmed by EPR spectra: $2.46 \cdot 10^{23}$ spin/g for Ni/C NC and $1.2 \cdot 10^{17}$ spin/g for Cu/C NC.

The presence of unpaired electrons on carbon hells of metal/carbon NCs gives the possibility to the interactions of these mesoparticles with electrofilled chemical substances. Below the mesoparticles interaction reactions with the compounds containing p, d elements are considered. The processes include the reactions without the elements OSs change and reactions with the change of elements OSs.

10.3 INTERFERENCE PHENOMENON AT METAL/CARBON NANOCOMPOSITES (NCS) MODIFICATION BY MEANS OF MESOPARTICLES INTERACTION WITH SUBSTANCES CONTAINING P D ELEMENTS

The interaction of metal/carbon nanocomposite (mesoparticle) with polyethylene polyamine (PEPA) and also ammonium iodide (AmI) is studied by IR and x-ray photoelectron spectroscopy. In these cases according to spectra, the nitrogen and iodine addition to carbon shell without change of OSs as well as the absence of metal atomic magnetic moment changes in mesoparticles is observed.

The modification processes of Cu/C NC is carried out by means of the grinding of mesoparticles with reagents (PEPA or AmI) in mechanical mortar longer than 3 minutes.

It's possible to suppose the following stages:

- Mechanical action on the mixture of mesoparticle and reagent initiates unpaired electrons on mesoparticle carbon shell and negative charge quants begin to move in side to positive charged atom of reagent.
- Then the increasing of reagent polarization with the negative charge quants flow formation proceeds, and farther the negative charge quants move to opposite flow of mesoparticle negative charge flow.
- Then the interference takes place and the following chemical bonds as C-N and C-I bonds are formed.
- In parallels the hydrogen from reagent (for instance, PEPA) is added to carbine bonds. Therefore, at the amine group interaction with mesoparticle carbon shell the formation of C=N bond is possible.

The proposed scheme is confirmed by the results of C1s and N1s spectra (Figures 10.4 and 10.5).

In C1s spectrum of Cu/C nanocomposite modified by PEPA the following components as:

C-H (285 eV), C-N (286.4 eV) and C=N (289 eV) are found. It's necessary to note that the peak at 289 eV has the small intensity. This fact is explained by the small quantity of NH_2 groups in PEPA because these groups are the ultimate groups in oligomer.

The presence of C-N (398.8 eV) and N-H (397 eV) bonds is observed in N1s spectrum of Cu/C nanocomposite.

The results of C1s and N1s spectra are confirmed the data of IR spectroscopy for samples modified by PEPA. In IR spectra the additional peaks at 1075 and 1268 cm^{-1} are appeared. These peaks can be relevant to C-N bonds in nitrogen-containing compounds including bonds in the adsorbed PEPA on the surface of Cu/C NC.

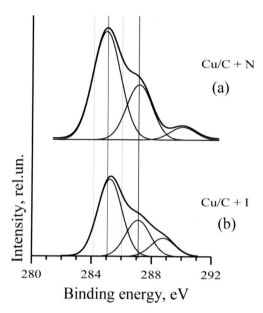

FIGURE 10.4 C1s spectra of Cu/C nanocomposite modified by PEPA (a) and by AmI (b).

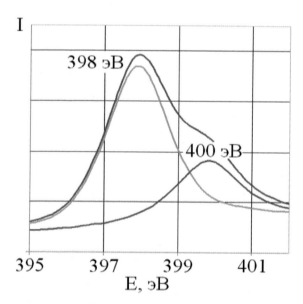

FIGURE 10.5 N1s spectra of Cu/C nanocomposite modified by PEPA.

10.4 ANNIHILATION PHENOMENON AT MESOPARTICLES MODIFICATION BY MEANS OF REDUCTION OXIDATION PROCESSES WITH PARTICIPATING IN THE SUBSTANCES CONTAINING P, D ELEMENTS

The annihilation phenomenon is possible at the reduction-oxidation (redox) processes when the positive charged atoms of oxidizer are linked with electron waves of reducer. If the reducer is the metal/carbon nanocomposite (mesoparticles) then the reduction of oxidized reagents such as ammonium phosphate (APPh), silica (SiO_2), metal oxides is stimulated by unpaired electrons which are found on carbon shell of correspondent mesoparticles. In these cases, the metal atomic magnetic moment growth in mesoparticles is discovered because of the shift of electrons on more high energetic levels. The electron transport in mesoparticles is explained by the creation of direct electromagnetic flow arised at annihilation or, in other words, at the interaction of positive charges quants and negative charges quants.

  The investigation of copper- and nickel/carbon NCs interaction with the following reagents as APPh, SiO_2, aluminum oxide, iron oxide, nickel oxide, and copper oxide is carried out. In correspondent reactions, the element reduction for reagents and nickel or copper atomic magnetic moments growth in mesoparticles takes place. In Table 10.2 the examples of metal atomic magnetic moments changes for mesoparticles modified by APPh or silica after the mechanochemical modification processes are given.

TABLE 10.2 The Values of Copper (Nickel) Atomic Magnetic Moments in the Interaction Products for Systems: Cu/C NC-APPh (or SiO_2) and Ni/C NC-APPh (or SiO_2)

Systems Cu/C NC-Substances	μ_{cu}	Systems Ni/C NC-Substance	μ_{Ni}
Cu/C NC-silica	3.0	Ni/C NC-silica	4.0
Cu/C NC-APPh	2.0	Ni/C NC-APPh	3.0
Cu/C NC-APPh, relation 1:0.5	4.2		

  The presence of unpaired electrons on mesoparticles carbon shells in above systems is determined by means of electron paramagnetic resonance (EPR) (Table 10.3).

  The metal atomic magnetic moment growth proceeds owing to the redox processes with above chemical compounds. In papers [6, 7] it's shown that the reduction reactions of phosphorus and silicon from correspondent substances at the interaction on the interphase boundary with mesoparticles are realized (Figures 10.6 and 10.7).

TABLE 10.3 The Unpaired Electron Values (from EPR Spectra) for Systems "Cu/C NC-Silica" and "Cu/C NC-APPh" (Relation 1:1) in Comparison with Mesoparticle Cu/C NC

Substance	Quantity of Unpaired Electrons, spin/g
Cu/C nanocomposite	1.2×10^{17}
system "Cu/C NC-SiO$_2$"	3.4×10^{19}
system "Cu/C NC-APPh"	2.8×10^{18}

According to Figure 10.4, the phosphorus reduction process proceeds more full at the relation 1:0.5 (with the decreasing of APPH layer) that is explained by the increasing of electron quantization.

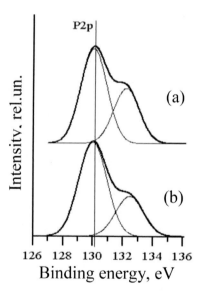

FIGURE 10.6 P2p spectra for systems: a. Cu/C NC-APPh 1:1; b. Cu/C NC-APPh 1:0.5.

The peak corresponding to energy EP2p equaled to 132.5 eV is attributed to C=N bond [8]. This fact can be explained by the interference which is increased because of the direct electromagnetic field formed at the annihilation. It's possible that the reduced phosphorus is found between carbon fibers of mesoparticle shell.

From Figure 10.5 the next conclusion can be made: the silicon reduction process is realized on 50% at the relations 1:0.5 and 1:1. This result may be explained by the decreasing of electron quantization in layers containing Si-O bonds.

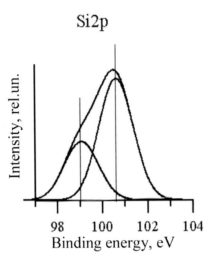

Si2p

FIGURE 10.7   Si2p spectrum of system "Cu/C-SiO$_2$" in relation 1:1 (E$_{Si2p}$=100.7 eV (SiO); E$_{Si2p}$=99eV (Si).).

Below the results of investigations by means of x-ray electron spectroscopy for Ni/C NC interactions with oxides of such metals as aluminum, iron, nickel, and copper are given in Tables 10.4–10.6 and on Figures 10.6 and 10.7.

The relations of NCs (mesoparticles) to above oxides are changed from 1:1 to 1:0.2 depending on the qualitative spectra obtaining, For example, the relations of 1:1 and 1:0.5 for system "Ni/C NC-Al$_2$O$_3$" leads to full mask of mesoparticle. Therefore, the quantity of aluminum oxide is decreased to the relation 1:0, 2. In accordance with Al3s spectra Aluminum is completely reduced during the modification process, and nickel atomic magnetic moment is increased to 4.8μ$_B$ (Table 10.4).

In this case, the reduction process is related to not only aluminum oxide but also to nickel oxide from metal cluster of mesoparticle (Ni/C NC). Therefore, the reduction processes are stipulated by the electron transport from carbon shell of mesoparticle in direction to Al^{+3} and Ni^{+2} of atoms in correspondent oxides.

TABLE 10.4   Parameters of Multiple Splintering Me3s Spectra in System $Ni/C + Al_2O_3$

Sample	I_2/I_1 (Ni)	Δ_{Ni}, eV	$\mu_{Ni}, \mu_{Б}$
Ni/C	0.3	3.0	1.8
Ni/C + Al_2O_3=1:0.2	0.9	1.0	4.8

I_2/I_1 – the relation of multiple splintering lines maximums intensities;
Δ – Energetic distance between the multiple splintering maximums in Me3s spectra.

The next researches with using of x-ray photoelectron spectroscopy are carried out for the following systems: Ni/C NC-Fe_2O_3 and Ni/C NC-CuO (Table 10.5).

TABLE 10.5   Parameters of Multiple Splintering Me3s Spectra in Systems "Ni/C NC-Fe_2O_3" and "Ni/C NC-CuO"

Sample	I_2/I_1 (Ni)	Δ_{Ni}, eV	$\mu_{Ni}\,\mu_{Б}$	$(I_2/I_1)_{Fe}$	Δ_{Fe}, eV	$\mu_{Fe}, \mu_{Б}$
Ni/C NC + (Fe_2O_3)	0.5	2.4	2.5	0.6	2.0	3.0
Ni/C NC + CuO	0.5	3.0	2.5	0.4	3.0	2.0

I_2/I_1 – the relation of multiple splintering lines maximums intensities;
Δ – Energetic distance between the multiple splintering maximums in Me3s spectra.

It's necessary to note, that the iron oxide is sufficient easily transformed in Fe_3O_4 (magnetite). At the heating, the farther reduction of iron with its implantation within the carbon shell is observed.

When copper oxide (CuO) is added to Ni/C nanocomposite (mesoparticle) in the relation 1:1 at the grinding, the relative intensity of multiple maximum for Ni3s and Cu3s spectra is correlated with the atomic magnetic moments values (Table 10.5). C1s spectrum for system "NiC NC-CuO" consists from three components: C-C (sp^2 hybridization) 284 eV; C-H bond 285 eV; C-C (sp^3 hybridization) 286 eV. The results of multiple splintering in Ni3s spectra for systems "Ni/C NC-CuO" and "Ni/C NC-NiO" are closed (Tables 10.5 and 10.6).

X-ray photoelectron spectroscopy investigations of Ni/C mesoparticles modified by nickel oxide are carried out for the following systems: a) Ni/C NC-NiO, relation 1:1; b) Ni/C NC-NiO, relation 1:0.5. The researches proceed without the sample heating and with the heating (300°C) of samples. When the studies are realized without heating the intensity of maximum Ni2p line corresponds to binding energy equaled to 855 eV (NiO). However, at the heating to 300°C, the intensity maximum for this line is attributed on binding energy to 852 eV (Ni). Below the Ni3s spectra for a) and b) systems without heating are shown (Figure 10.8).

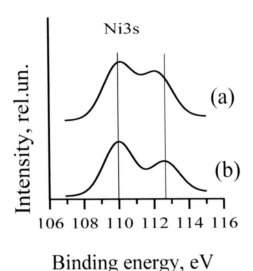

FIGURE 10.8   Ni3s spectra for systems: (a) NiC NC-NiO (1:0.5); (b) NiC NC-NiO (1:1).

The oxygen content is more much decreased at the sample heating. The C1s spectra (Figure 10.9) consist from three components which take place [2] in carbon nanostructures: 285 eV (C-H bond); 284 eV (sp^2 hybridization); 286 eV (sp^3 hybridization).

According to C1s spectra for system "Ni/C NC-NiO" (relation 1:1) the maximum characteristic of sp^2 hybridization is only remained at the sample heating to 300°C (Figure 10.7b), and the C-H component intensity for the sample with relation 1:0.5 is decreased as well as the correspondent intensity at the relation 1:1 becomes equaled to zero. These changes are accompanied by correspondent changes of nickel atomic magnetic moments:

- For system of content 1 Ni/C NC–0.5 NiO the Ni atomic magnetic moment increasing corresponds to 1.2 μ_B;
- For content system 1 Ni/C NC–1NiO analogous increasing value equals to 2.2 μ_B (Table 10.6).

It's possible that the increasing of Nickel atomic magnetic moments proceeds because of the growth of atomic magnetic moments both mesoparticle cluster metal and modifier (NiO) metal.

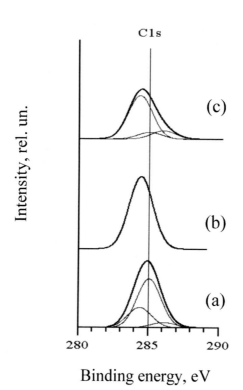

FIGURE 10.9   C1s spectra: (a) surface without heating; (b) Ni/C-NiO, 1:1 (300°C); (c) Ni/C-NiO, 1:0.5 (300°).

TABLE 10.6   Parameters of Multiple Splintering Me3s Spectra in Ni/C Nanocomposite (Ni/C NC) and in Systems, Containing Ni/C NC-NiO

Sample	I_2/I_{1Ni}	Δ_{Ni}, eV	$\mu_{Ni}, \mu_{Б}$
Ni/C NC	0.32	3.0	1.8
Ni/C NC+ NiO= 1:0.5	0.6	2.6	3.0
Ni/C NC+ NiO= 1:1	0.8	2.2	4.0

I_2/I_1 – the relation of multiple splintering lines maximums intensities;
Δ – Energetic distance between the multiple splintering maximums in Me3s spectra.

The distance decrease between the maximums of multiple splintering in Ni3s spectra for systems shows on the increase the hybridization of d electrons in nickel atoms that is the reason of nickel atomic magnetic moment growth.

10.5 CONCLUSION

On the basis of ideas about charge quantization, and also about the chemical bond formation because of the interference of electron waves as well as the hypothesis about annihilation appeared at the interactions of positive and negative charges quants the analysis of modification processes for the reactions between copper or Ni/C NC (mesoparticles) and substances containing such elements as nitrogen, iodine, phosphorus, silicon, aluminum, iron, copper, and nickel is carried out. In the middle of its substances as PEPA, AmI, ammonium polyphosphate (APP), silica (SiO_2), aluminum oxide, iron oxide, nickel oxide, and copper oxide are used. In the case, when PEPA and AmI are applied, the connection reactions takes place. As the main method in the investigations x-ray, photoelectron spectroscopy is applied. Owing to this method, the chemical bonds formation between mesoparticles carbon shells and nitrogen or iodine is found because of the interference of negative charge quants in the interaction processes of mesoparticles with PEPA or AmI. At the interactions of PEPA with mesoparticles, the C-N bond formation is explained by the interference of negative charges quants. In these cases the modifiers reduction reactions takes place. The energy which is evolved at this can be used in chemical reactions including connection to mesoparticle. On the base of studies results, the hypothesis of possibility of two phenomena (interference and annihilation) in redox processes is proposed. The correspondence of results obtained by the methods of x-ray photoelectron spectroscopy and EPR is given. The changes for metal atomic magnetic moments from the number of electrons which participated in redox process are observed.

Thus, the magnetic properties regulation possibilities for the obtaining of mesopaticles with more much atomic magnetic moments are discovered on the examples, mechanochemical modification of metal/carbon NCs by the substances containing positive charged atoms with a high positive charge.

KEYWORDS

- **annihilation**
- **charges quantization**
- **mesoparticles**
- **metal atomic magnetic moment**
- **metal/carbon nanocomposites**
- **redox processes**

REFERENCES

1. Kodolov, V. I., & Trineeva, V. V., (2017). New scientific trend: Chemical mesoscopics. *Chemical Physics and Mesoscopics*, *19*(3), 454–465.
2. Yavorskiy, B. M., & Detlaph, A. A., (1983). *Directory on Physics* (pp. 519, 848). M.: Publ. "Science," 1965, Kuhling H. reference book on physics. M.: Publ. "World."
3. Rudenberg, K., (1964). *Physical Nature of Chemical Bond* (p. 162). M.: Publ. "World."
4. Kodolov, V. I., Trineeva, V. V., Kopylova, A. A., et al., (2017). Mechanochemical modification of metal/carbon nanocomposites. *Chemical Physics and Mesoscopics*, *19*(4). 569–580.
5. Shabanova, I. N., Kodolov, V. I., Terebova, N. S., & Trineeva, V. V., (2012). *X-Ray Electron Spectroscopy in Investigations of Metal/Carbon Nanosystems and Nanostructured Materials* (p. 252). M.-Izhevsk: Publ. "Udmurt University."
6. Kodolov, V. I., Trineeva, V. V., Terebova, N. S., et al., (2018). Changes of electron structure and magnetic characteristics of modified copper/carbon nanocomposites. *Chemical Physics and Mesoscopics*, *20*(1), 72–79.
7. Kopylova, A. A., & Kodolov, V. V., (2014). Investigation of copper/carbon nano-composite interaction with Si atoms from silicon compounds. *Chemical Physics and Mesoscopics*, *16*(4), 556–560.
8. Wang, J. Q., Wu, W. M., & Feng, D. M., (1992). *The Introduction to Electron Spectroscopy (XPS/XAES/UPS)* (p. 640). Beijing: National Definite Industry Press.

CHAPTER 11

Magneto-Electric Properties of Sodium Potassium Lithium Niobate-Ni/Co Ferrite Nanocomposites

R. RAKHIKRISHNA[1] and J. PHILIP[2]

[1]Department of Instrumentation, Cochin University of Science and Technology, Cochin–682022, India

[2]Amal Jyothi College of Engineering, Kanjirappally, Kottayam–686518, India, E-mail: jp@cusat.ac.in

ABSTRACT

The current chapter deals with selected lead-free multiferroic magneto-electric composites. Lead-based piezoelectric materials show incompetently high value of the piezoelectric coefficient. So they are inevitable in electronic industries nowadays. However, lead is a highly toxic and environmentally hazardous material. So there is a growing interest in the development of lead-free piezoelectrics having good efficiency.

11.1 INTRODUCTION

Correlated systems maintain some of the characteristic properties of their constituent atoms in isolation. Magnetism is such a characteristic exhibited by such systems [1, 2]. Multiferroics generally come under the category of strongly correlated systems. Multiferroics possess more than one ferroic order such as ferroelectricity, ferroelasticity, ferromagnetism, etc. and are smart to perform more than one task at a time. Coexistence of ferroelectric and ferromagnetic orders in a material can result in magneto-electric behavior. Magneto-electric materials have mutual control in electric and magnetic realm. Such materials can have revolutionary applications in manufacturing memory devices, actuators, sensors, spintronic devices, etc. If electric polarization can be controlled by external magnetic fields and spins by applied voltage, new devices with varieties of functions can be designed. Extensive researches have been conducted on magneto-electric effect because of the obvious potential of coupling between magnetic and electric properties, which is the basis for their technical applications.

The story of such materials begins in 1894, when Curie demonstrated induction of electric polarization in some asymmetric molecular bodies in magnetic field [3]. A number of materials having multiferroic properties have been discovered. There are two types of magneto-electric materials, single phase, and composites. It is rare to obtain both electric and magnetic properties in a single material because of their mutually exclusive nature. Ferroelectricity, having permanent polarization is the result of the relative shift of positive and negative ions, usually seen in material with empty d orbitals. But existence of partially filled d orbitals is an essential condition for the occurrence of magnetism in a material, because magnetism is the relative ordering of spins of electrons in incomplete ionic shells [4–6]. Even though it is difficult to obtain single-phase magneto-electric materials [4], very few of them exist in nature due to adequate structural features that permits ferroelectric type ionic movements, superexchange type magnetic interactions, symmetry conditions, dipolar interactions and Zeeman energy [7, 8]. Bur single-phase materials usually have very low room temperature magneto-electric properties. So the artificial composite materials are evolved, which have an advantage of greater design flexibility and multi-functionality at room temperature. In this chapter, we are dealing with selected lead-free multiferroic magneto-electric composites. So there is a growing interest in the development of lead-free piezoelectrics having good efficiency.

In magneto-electric composites [9], the magneto-electric property originates from the cross-coupling of piezoelectric and piezomagnetic properties, because they comprise of piezoelectric and magnetostrictive phases (see Figure 11.1). The efficiency of material as a magneto-electric one can be determined by its magneto-electric coefficient. Theoretically, the magneto-electric property is the product-property of piezoelectric and magnetostrictive properties explained in Section 2. But, in practice, the efficiency of such magneto-electric materials also depends on various other parameters like mechanical coupling and interfacial effects. The changes in properties of a composite also depend on parameters like particle size of individual phases, interfacial effects, poling, etc.

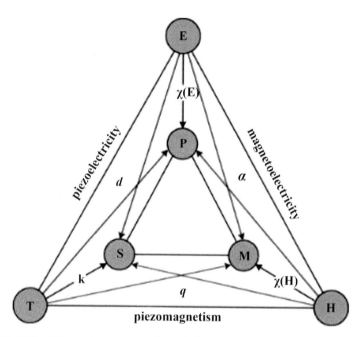

FIGURE 11.1 Schematic representation of the relationship between ferroelectricity (polarization P and electric field E), ferromagnetism (magnetization M and magnetic field H), and ferroelasticity (strain S and stress T), piezoelectricity (stress T and electric field E), piezomagnetism (stress T and magnetic field H) and their coupling and mutual control in condensed matter represent the cores of multiferroic and magneto-electric phenomena.

11.2 CONCEPT OF PRODUCT-PROPERTY

The magneto-electric effect in composite materials is a product tensor property [10]. The product property was proposed by Van Suchtlen in 1972 [11]. The cross interaction of electric and magnetic phases in the composite is explained by this property. The magneto-electric interactions in composites are strain mediated and the magneto-electric coefficient is the product of magnetostrictive coefficient (magnetic/mechanical in the magnetic phase) and piezoelectric coefficient (mechanical/electrical in piezoelectric phase).

$$\text{Direct M-E effect} = (\frac{\text{magnetic}}{\text{mechanical}}) \times (\frac{\text{mechanical}}{\text{electric}})$$

$$\text{Converse M-E effect} = (\frac{\text{electric}}{\text{mechanical}}) \times (\frac{\text{mechanical}}{\text{magnetic}})$$

The strain developed in the magnetostrictive phase upon applying a magnetic field exerts a stress on the piezoelectric phase. This results in an induced polarization on the strained piezoelectric phase. The converse effect, that is the induction of magnetization by applying an electric field, is also possible. From this concept, the magneto-electric effect and thus the magneto-electric coefficient of the composite is high only on combining materials with high piezoelectric coefficient d_{33} and magnetostrictive coefficient q_{11}. In addition to this, the magneto-electric effect also depends on the composite microstructure and coupling interaction across magnetic piezoelectric interfaces.

11.3   0–3 PARTICULATE FERRITE COMPOSITES

One can prepare composites with different connectivity schemes. The concept of phase connectivity of composites was introduced by Newnham et al. [12]. Multiferroic magneto-electric composites of different connectivity schemes have been developed [13] and more common connectivity schemes like 0–3 particulate, 2–2 laminate and 1–3 fiber-rod type are shown in Figure 11.2. 0–3 Particulate composite means single-phase particles having zero dimensions are embedded in a 3D matrix of another phase.

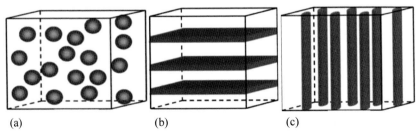

(a)   (b)   (c)

FIGURE 11.2   Bulk composites with the three common connectivity schemes: (a) 0–3 particulate composite, (b) 2–2 laminate composite, (c)1–3 fiber/rod composite. Each number represents the connectivity of the phase.

In the history of magneto-electric composites, a high value of magneto-electric coefficient (0.13 V/cm-Oe) at room temperature was obtained in composites of $BaTiO_3$-$COFe_2O_4$ prepared by unidirectional solidification soon after the product-property was proposed by scientists at Philips Laboratory [13–16]. Composites having magneto-electric coefficient 100 times larger than that of a single-phase have been prepared. Unidirectional solidification is a complex route and requires critical control over composition and processing.

Normal sintering is easier and cost-effective than solidification to synthesize eutectic composites. In the early 1990s, many authors prepared magneto-electric composites by conventional normal sintering route [13]. Even though composites prepared by normal sintering show low magneto-electric coefficient compared to prior, many advantages are there compared to in situ composites in combining phases. They provide an opportunity to combine phases with widely different crystal structures. Moreover, the molar ratio of phases, grain size of each phase, and sintering temperature are easily controllable [17].

11.4 MAGNETO-ELECTRIC COEFFICIENT (*A*)

In order to study various parameters that depend on the properties of magneto-electric composites, we have to look into the kind of coupling and interaction that lead to magneto-electric phenomenon. The linear approximation of electric-mechanical-magnetic coupling in composites can be expressed by the tensor notation [18]:

$$S_i = k_{ij}T_j + d_{ki}E_k + q_{ki}H_k \tag{1}$$

$$D_k = d_{ki}T_i + \varepsilon_{kn}E_n + \alpha_{kn}H_n \tag{2}$$

$$B_k = q_{ki}T_i + \alpha_{kn}E_n + \mu_{kn}H_n \tag{3}$$

where, H = magnetic field intensity; B = magnetic flux density; k = mechanical compliance (at zero magnetic field); μ = magnetic permeability at constant stress (zero mechanical stress); q = piezomagnetic constant; T =stress factor; S =strain factor; E =vector electric field;

ε = permittivity at constant stress (at zero mechanical stress); d = piezo-electric strain constant;

D = vector of electric displacement; and α = magneto-electric coefficient.

The magneto-electric coefficient is the parameter that determines the efficiency of a material to work as a magneto-electric transducer. In composites, macroscopic magneto-electric effect is the combination of two types of materials property such as magnetostriction and piezoelectricity.

$$\Delta P = \alpha \, \Delta H \text{ (direct magneto-electric effect)}$$

$$\Delta E = \alpha_E \, \Delta H \text{ (converse magneto-electric effect)}$$

where; α and α_E are the magneto-electric coefficient and magneto-electric voltage coefficient respectively.

$$\Delta M = \alpha \, \Delta E$$

where; α is the magneto-electric magnetization coefficient.

Thermodynamically, magneto-electric effect can be understood within the Landau theory framework, approached by the expansion of free energy for a magneto-electric system, i.e., [13, 17]

$$F(E,H) = F_0 - P_i^s E i - M_i^s H i - 1/2\varepsilon_0\varepsilon_{ij}E_iE_j - 1/2\mu_0\mu_{ij}H_iH_j$$
$$-\alpha_{ij}E_iH_j - \beta_{ijk}E_iH_jH_k - \gamma_{ijk}H_iE_jE_k - \ldots \tag{1}$$

F_0 is the ground state free energy, i, j, k represent three spatial coordinates, E_i, and H_i are the components of the electric field E and magnetic field H, respectively, P^s, and M^s are the spontaneous polarization and magnetization respectively, ε_0 and μ_0 are the dielectric and magnetic susceptibilities of vacuum, ε_{ij} and μ_{ij} are the second-order tensors of dielectric and magnetic susceptibilities, β_{ijk} and γ_{ijk} are the third-order tensor coefficients and, α_{ij} is the components of tensor α which is the linear M-E effect and corresponds to the induction of polarization by a magnetic field or a magnetization by an electric field. Then $\alpha = \varepsilon_0 \cdot \varepsilon_r \cdot \alpha_E$, where ε_r is the relative permittivity.

The magneto-electric effect is usually measured under two distinctly different conditions which refer to electric polarization induced by an applied magnetic field or magnetization induced by an external electric field. So it is more convenient to write the above equation in terms of polarization P and magnetization M [18, 19].

$$P_i(E,H) = -\frac{\partial F(E,H)}{\partial Ei} = P_i^s + 1/2\varepsilon_0\varepsilon_{ij}E_j + \alpha_{ij}H_j + \beta_{ijk}H_jH_k + \ldots$$

Polarization in magnetic field $E=0$,

$$P_i(H) = \alpha_{ij}H_j + \beta_{ijk}H_jH_k + \ldots$$

$$M_i(E,H) = -\frac{\partial F(E,H)}{\partial Hi} = M_i^s + 1/2\mu_0\mu_{ij}H_j + \alpha_{ij}E_i + \gamma_{ijk}E_jE_k + ...$$

Magnetization in electric field $H=0$

$$M_i(E) = \alpha_{ij}E_i + \gamma_{ijk}E_jE_k + ...$$

Theoretically, magneto-electric output in composites should increase with the ferrite content and it would be maximum for the composite with 50% ferrite and 50% ferroelectric provided that there is no depolarizing field.

Zhai et al. [20], while experimenting with PZT/NiFe$_2$O$_4$ system, reported that α_{E33} of this composite decline after a peak value with increase in volume fraction of ferrite. This is because the high concentration of ferrite phase makes it difficult to polarize the composite. High ferrite content results in low piezoelectric coefficient value and the charge developed on piezo phase leaks through the low resistance ferrite path.

11.5 SYNTHESIS AND CHARACTERIZATION OF (Na$_{0.5}$K$_{0.5}$)$_{0.94}$ Li$_{0.06}$ NbO$_3$-Ni/Co Fe$_2$O$_4$ NANOCOMPOSITES (NCs)

Normal sintering method is adopted in our work to prepare (Na$_{0.5}$K$_{0.5}$)$_{0.94}$ Li$_{0.06}$ NbO$_3$ – Ni/Co Fe$_2$O$_4$ nanocomposites (NCs). Sintering has greater design flexibility and controllability on grain growth and other physical parameters [17]. In order to fabricate magneto-electric 0–3 composites, physical mixing of individual phases has been done followed by high-temperature sintering. The ferrite nanoparticles (NPs) are synthesized by wet chemical or co-precipitation route. For this, aqueous solutions of 0.4 molar ferric chloride and 0.2 molar metal Ni/Co chloride solutions are mixed and stirred well in an alkaline pH. In order to reduce agglomeration, two or three drops of a surfactant, oleic acid, are added and allowed to mix in a temperature of about 90°C. The precipitate obtained is washed well with deionized water and acetone to remove unreacted salts, extra-base, and surfactant. The calcination was carried out at 600°C for 10 hours [21]. Bulk piezoelectric phase is prepared by conventional solid-state reaction route. Stoichiometric ratios of Na$_2$CO$_3$, K$_2$CO$_3$, Li$_2$CO$_3$, Nb$_2$O$_5$ are mixed using mortar and pestle in ethanol medium and the powder so obtained is calcined at 850°C for 10 hours to get the piezoelectric phase [22].

The powder compacts of mixture of phases are prepared by adding binder, polyvinyl alcohol (PVA), and pressing the solid solution under a pressure of 7 Tons. The composites are obtained by sintering the powder compacts at 1000°C for 4 hours.

Physical characterizations of the composites are done with X-ray powder diffractometer (Bruker, Model AXS D8 Advance) with incident Cu-Kα radiation of λ=1.54 A° and Fourier transform infrared spectroscopy (FTIR) (Thermo Nicolet, Avatar 370). Piezoelectric measurements are carried out by Piezotest d33 Piezo Meter System (Model PM300). Magnetic measurements on the composites are carried out with a vibrating sample magnetometer (VSM) (EG and PAR, Model 4500). Dielectric measurements of the samples are performed using an Impedance analyzer (Hioki, Model IM 3570). Magneto-dielectric (MD) measurements are done by keeping the sample between the pole pieces of an electromagnet. Magneto-electric measurement set up which uses dynamic method to measure the magneto-electric property is used to determine the magneto-electric coefficient of the samples. The magneto-electric voltage co-efficient, which is the parameter that determines the efficiency of the material, can be measured as the voltage induced per unit thickness with applied *ac* magnetic field on a parallel constant *dc* bias field.

11.6 DYNAMIC MEASUREMENT TO DETERMINE THE MAGNETO-ELECTRIC COEFFICIENT

Dynamic measurement of non-linear magneto-electric effect in a multiferroic magneto-electric composite requires application of parallel *ac* and *dc* magnetic fields. The magneto-electric property can be measured as a variation of $\frac{dE}{dH}$, which is the magneto-electric coefficient α_E as a function of *dc* magnetic bias field. For this, an electro-magnet can be used for the bias field. This *dc* magnetic field can be measured using a Tesla meter. The coefficient can be measured directly as a response of the sample to an *ac* magnetic signal superimposed on the *dc* bias field. Both *ac* and *dc* fields are applied parallel to the sample axis. A Helmholtz coil can be used to generate *ac* magnetic field. A signal generator is required to drive this Helmholtz coil. On applying the fields, charges are generated on the piezo phase of the sample and this can be measured using a charge amplifier. This charge amplifier is so designed as to convert a charge signal from piezoelectric transducer into proportional output voltage. An oscilloscope is used to measure this output voltage from the charge amplifier. This output voltage can be represented as an electric charge from the piezo phase under short circuit conditions. The output voltage from the charge amplifier is obtained from the basic relation $V = \dfrac{Q}{C}$ at 1 kHz [23]. A schematic diagram of the measurement set up is shown in Figure 11.3.

$$\text{Magneto-electric voltage coefficient, } \alpha_E = \frac{dE}{dH} = \frac{V}{(t \, x \, Hac)}$$

where; t is the thickness of the sample, H_{ac} is the *ac* magnetic field, and V is the output voltage from the oscilloscope. From these, the magneto-electric voltage coefficient of the sample can be calculated.

Zhai et al. [20] measured magnetic anisotropy of the sample by rotating the sample holder and thereby changing its direction of orientation with respect to the applied magnetic field.

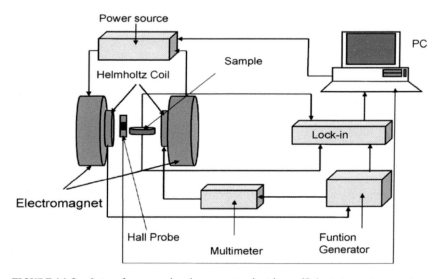

FIGURE 11.3   Set up for measuring the magneto-electric coefficient at room temperature.

11.7   POLING OF THE SAMPLE

Before carrying out magneto-electric measurements, samples have to be poled both electrically and magnetically. For the electrical poling of ceramic pellets, contact poling method is efficient to align electric dipoles inside the material. A *dc* electric field of 1–3 kV/mm is used in this work to pole the prepared samples. Poling voltage is applied at a temperature of 150°C for half an hour, followed by cooling at the same voltage. Electric poling of this material is difficult due to the low resistance ferrite phase present in the composite. Ferrite phase provides a low resistive leakage path for the charged particles through the sample. In order to get effective poling in magneto-electric composites, samples should be optimized first. Results on studies on different particulate composites support low ferrite content of about 20% as the optimum composition [24]. The problems due to charge leakage and distraction of the sample could be avoided by limiting the current through it with a series resistance. Before carrying out the MD and magneto-electric measurements, the samples are not only poled electrically but also magnetically by a *dc* magnetic field of field strength 1T for 30 minutes [25].

11.8   STRUCTURAL CHARACTERIZATION OF $(Na_{0.5}K_{0.5})_{0.94} Li_{0.06}$ NbO_3 – MFe_2O_4 BY X-RAY DIFFRACTION (XRD)

X-ray diffraction (XRD) is the most effective analytical technique used to identify the crystallographic structure and phases. When X-rays are applied to the powder samples, the diffraction intensity is assumed to be the sum of X-rays reflected from all the fine grains and the peaks are attributed to the Miller indices of the sample [26]. Lead-free magneto-electric NCs of sodium potassium lithium niobate-nickel/cobalt ferrite (xNKLN-(1-x)MFO), with different molar weight percentages, have been characterized using XRD. Figure 11.4 shows the XRD spectra of the composites prepared. Panalytical X'Pert PRO high-resolution XRD with incident Cu-Kα radiation of λ=1.54 A° is used for the structural characterizations of the composite. The XRD of both phases are compared with the results reported in literature and the spectra match well with them [21, 22]. In order to identify the (*h k l*) planes, Pawley method has been used. The NKLN perovskite belongs to the space group symmetry *Cm2m*. The lattice parameters obtained are a = 5.637 A°, b = 5.669 A° and c = 3.945 A°. Ferrite phase has face-centered cubic spinel structure. Spectrum can also be verified using standard ICSD file. The symmetry group of ferrites is identified as *Fd3m*. In the XRD spectra of composites, the peaks of individual phases are clearly visible and do not contain any additional peaks. This reveals that there is no chemical reaction between individual phases of the composite.

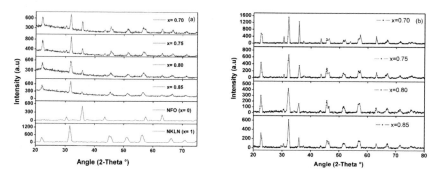

FIGURE 11.4 X-ray powder diffraction spectrum of $(Na_{0.5}K_{0.5})_{0.94} Li_{0.06} NbO_3-MFe_2O_4$ (a) M is nickel (b) M is cobalt.

11.9   FOURIER TRANSFORM INFRARED SPECTROSCOPY (FTIR) OF $(Na_{0.5}K_{0.5})_{0.94} Li_{0.06} NbO_3-MFe_2O_4$

Fourier transform infrared (FTIR) spectroscopy is generally used for chemical analysis. They are also useful in identifying typical molecular structures. Figure 11.5 shows the recorded FTIR spectra of the composite samples in the wavenumber range 4000 cm^{-1} to 400 cm^{-1}. Magneto-electric composite is a combination of different types of metal oxides. So the FTIR spectra will contain vibrational peaks of metal oxides. It is reported that in ferrites, the stretching vibration of the tetrahedral metal-oxygen bond in the range 600–550 cm^{-1} and octahedral metal-oxygen bond in the range 450–385 cm^{-1} provide confirmation for the formation of such bonds [27]. Tetrahedral metal-oxygen bond vibration found in our measurement range confirms the formation of metal oxides, which are clearly visible in Figure 11.5.

11.10   MAGNETIC CHARACTERIZATION BY VIBRATION SAMPLE MAGNETOMETRY

M-H hysteresis measurement can be used to study the magnetic properties of composites. Vibration Sample Magnetometer (VSM) is generally used to plot the magnetic hysteresis. A magneto-electric composite will perform efficiently only when the electric and magnetic properties are well maintained. The magnetic properties of such composites are due to the presence of magnetic ferrite phase. The individual ferrite grains act as the centers of magnetization and the saturation magnetization of the composites is the vector sum of all these individual contributions. The magnetic content increases with ferrite content and result in increase of net magnetization. But, the sample has to optimize the composition by taking care of the cross-linked chains of magnetic structures. This is because these low resistive ferrite chains will lead to electrical leakage paths that inhibit domain growth while poling, resulting in low piezoelectric and magneto-electric properties. On considering magnetic properties, the ferroelectric materials incorporated into the ferrite phase acts as pores in the presence of applied magnetic field and break the magnetic circuit. This will result in the decrease of these magnetic parameters with increasing ferroelectric concentration. The magnetic behavior of ferrites depends on the number of parameters like cation distribution, site preference energies, covalence of bonds, and the molecular field [20, 28, 29]. But in a composite system, interfacial effects are also expected play a role on the magnetic interactions due to change in the distribution of magnetic ions and spin orientations [30]. Figure 11.6 is the magnetic hysteresis curves of the composites prepared, which is the conformation of magnetism in the composite material. Thin hysteresis curve of nickel ferrites based composites indicates soft magnetic behavior compared to cobalt ferrite composites.

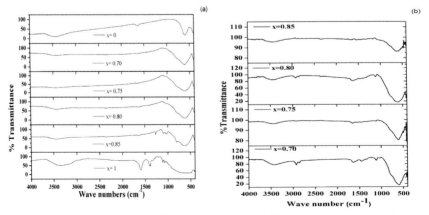

FIGURE 11.5   FT-IR spectra of $(Na_{0.5}K_{0.5})_{0.94} Li_{0.06} NbO_3 - MFe_2O_4$ composite showing metal-oxygen bond in which M is (a) nickel (b) cobalt.

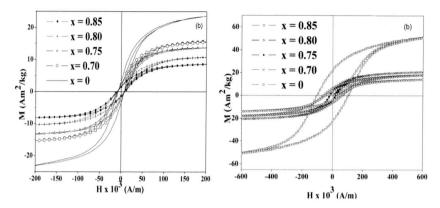

FIGURE 11.6 M-H hysteresis of $(Na_{0.5}K_{0.5})_{0.94} Li_{0.06} NbO_3-MFe_2O_4$ composite showing metal-oxygen bond in which M is (a) nickel (b) cobalt.

11.11 DIELECTRIC PROPERTIES OF $(Na_{0.5}K_{0.5})_{0.94} Li_{0.06}$ $NbO_3-MFe_2O_4$

Dielectric properties of magneto-electric materials as functions of external parameters like temperature and frequency of applied field have been widely studied by many authors [31–34]. At very low frequencies, there is a sharp fall in the frequency dependence of dielectric properties with increasing frequency. This sudden fall-off in dielectric constant with increase in frequency in the low-frequency region is attributed to Maxwell-Wagner type interfacial polarization [35, 36], in agreement with Koop's phenomenological theory [37]. Accordingly, the friction between dipoles increase. Due to friction, the dipoles dissipate energy in the form of heat, affecting the internal viscosity of the system; hence, there is a decrease in the dielectric constant. Space charge polarization due to in homogeneity in structure, like impurities, dislocations, porosity, and grain structure, are responsible for the large value of ε at low frequencies. This is attributed to the dipoles resulting from changes in valence states of cations and space charge polarization. Space charge polarization is governed by the number of space charge carriers and resistivity of the sample. The charge carriers, which take part in exchange, can be produced during sintering. In the case of composites, the ferroelectric region is surrounded by non-ferroelectric regions. This gives rise to interfacial polarization. In the high-frequency region, ε remains almost constant due to large inertia of electric dipoles, and ions follow up the fast variations in *ac* applied electric field. This is known as static value of dielectric constant [38].

Frequency dependence of room temperature dielectric parameters, from 100 kHz to 1 MHz, for one of the dense samples (with *x*= 0.85), for both nickel ferrite composite and cobalt ferrite composite, are plotted in Figures 11.7a and 11.7b respectively. In this frequency band, the dielectric parameters are almost constant or static. Studies on dielectric properties have been carried out before and after poling the samples, and are plotted in a single plot (Figure 11.7). Poled samples show resonance in the dielectric constant at different frequencies, and dielectric loss curve also shows sharp peaks at same frequencies. Poling align the electrical dipoles and thereby achieve domain growth in the ferroelectric phase. The reason for the occurrence of multiple resonances in dielectric curve after poling is due to the activation of different piezoelectric vibration modes. This comes from particulate microstructural features and random orientation of crystallites with respect to the applied field [39, 40]. Other dense samples also show similar behavior.

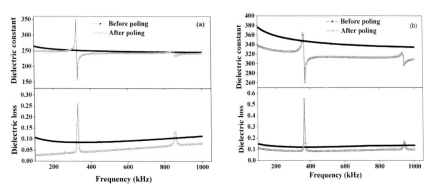

FIGURE 11.7   Frequency response of dielectric constant and loss factor before and after poling the sample $(Na_{0.5}K_{0.5})_{0.94} Li_{0.06} NbO_3–MFe_2O_4(x=0.85)$ where a) M is nickel and b) M is cobalt.

11.12   MAGNETO-DIELECTRIC (MD) OR MAGNETO-CAPACITANCE EFFECT

Interesting observation of the variation of dielectric constant as a function of applied magnetic field, or MD or magneto-capacitance effect, has been studied by us. The MD effect is due to variation of dielectric constant with applied stress that occurs due to piezomagnetic effect in the ferrite phase. MD coefficient should be proportional to ($\lambda k_m d\varepsilon/ds$), where λ is the magnetostriction coefficient, $d\varepsilon/ds$ is the rate of change of dielectric constant with induced stress due to applied magnetic field, and k_m is the mechanical coupling factor [29]. It is reported that the increases in stress will cause an increase in P_{max} and therefore ε decreases, making $d\varepsilon/ds$ negative. Therefore, MD will be positive for negative λ, and negative for positive λ. Some authors working on the polymeric composites also have reported MD behavior of the magneto-electric composites [31, 42].

The magnetic field dependent dielectric constant, MD, or magneto-capacitance coefficient is defined as [31, 43]

$$MD = \frac{\varepsilon_r(B) - \varepsilon_r(0)}{\varepsilon_r(0)} \times 100 \qquad (2)$$

In Figures 11.8(a) and 11.8(b), we plot the variations of dielectric constant with external magnetic field strengths at different frequencies for nickel ferrite and cobalt ferrite composites respectively. Before MD measurement, electrically poled samples are subjected to a *dc* magnetic field for magnetic poling. The dielectric constant of this material gets changed after magnetic poling. The dipoles inside the material are slightly realigned because the polarization of a magneto-electric material also depends upon external magnetic field.

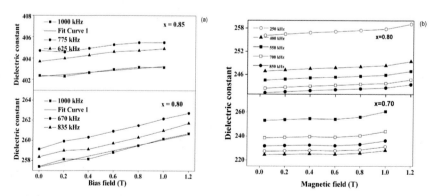

FIGURE 11.8 Variations of dielectric constants with external magnetic field at different frequencies of the sample $(Na_{0.5}K_{0.5})_{0.94} Li_{0.06} NbO_3 - MFe_2O_4$ (x=0.85) where a) M is nickel and b) M is cobalt.

The variations of MD (%) with external magnetic field at different frequencies for samples are shown in Figure 11.8. The values of MD (%) are low because of low magnetostriction coefficient λ with low ferrite content. To get a measurable linear MD (%) magnetostriction coefficient should be sufficiently large and mechanical coupling should be proper. Not all the prepared composites show such a linear MD effect. Maximum value of MD (%) obtained for the composite is only 1%, at a frequency of 1000 kHz, over a field variation of 1 T. These values of MD (%) are comparable to the results reported by Zhou et al. [44] on composites of $0.4CoFe_2O_4$-$0.6[0.948(K_{0.5}Na_{0.5}) NbO_3$-$0.052LiSbO_3]$.

11.13   MAGNETO-ELECTRIC VOLTAGE COEFFICIENT OF $(Na_{0.5}K_{0.5})_{0.94} Li_{0.06} NbO_3$ – MFe_2O_4 NANOCOMPOSITE

The efficiency of a magneto-electric material to convert electrical energy into magnetic energy and vice versa is determined by the magneto-electric voltage coefficient, α, given by:

$$\alpha = \left(\frac{dE}{dH_{ac}} \right)_{H_{dc}} \tag{5}$$

In magneto-electric composites, magneto-electric voltage coefficient is defined as the product property of individual phases, *viz.* piezoelectric and magnetostrictive phases and is mediated by strain [13]. In the experiments of Jigajeni et al [41], it is found that α increases with sintering temperature, attributable to the increasing grain size and the resulting improved magneto-mechanical coupling. The magnitude of α is known to be proportional to $(\lambda k_m d)/\varepsilon$, where λ is the magnetostriction coefficient, k_m is magneto-mechanical coefficient, d is the piezolelectric coefficient and ε is the dielectric constant. Therefore, the magneto-electric coefficient of the composite does not only depend on piezoelectric and magnetostriction coefficients but also the mole fractions of component phases, resistivity of phases and mechanical coupling between them. Porosity in the sample and increase in volume fraction of low resistive ferrite phase make effective poling difficult for composites. Porosity reduces mechanical coupling between phases, and increase in ferrite phase provides a leakage path. So, efficient magneto-electric conversion will be obtained only for composites with high enough density and sufficient ferrite content.

In Figure 11.9 we show the plot of the electric field developed across the sample against strength of applied *ac* magnetic field, keeping the bias *dc* magnetic field constant, for the dense and poled samples. From the slope of the plot the magneto-electric voltage coefficient α is obtained as:

$$\alpha = \frac{V}{\left(t \times H_{ac}\right)} \tag{6}$$

where; t is the thickness of sample, V is the output voltage and H_{ac} is the applied ac magnetic field in a constant *dc* magnetic field [24, 45]. The maximum values of the coefficient α obtained in $(Na_{0.5}K_{0.5})_{0.94}$ $Li_{0.06}$ NbO_3-$NiFe_2O_4$ is 0.024 *V/cm-Oe* for $x = 0.85$ [24] and is 0.0157 *V/cm-Oe* for $x = 0.80$ [46]. These are the highest values for α reported so far for any lead-free sample prepared by normal sintering method. The highest value reported so far for the magneto-electric coefficient for a lead-free sample prepared by simple solid-state reaction method is 0.013 *V/cm-Oe,* for the composite 0.5 $Ni_{0.3}Zn_{0.62}Cu_{0.08}Fe_2O_4$ + 0.5 $Pb(Fe_{0.5}Nb_{0.5})O_3$ [47].

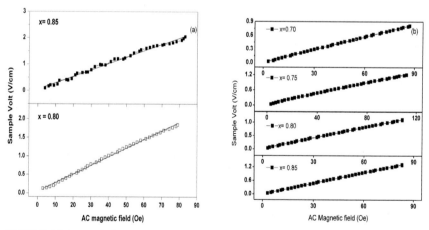

FIGURE 11.9   Variations of induced electric field in $(Na_{0.5}K_{0.5})_{0.94}$ $Li_{0.06}$ $NbO_3 - MFe_2O_4$ nanocomposite with applied ac magnetic field in a constant dc bias field where a) M is nickel and b) M is cobalt.

11.14   APPLICATIONS OF MAGNETO-ELECTRIC MATERIALS

The mutual interactions in electric and magnetic realms are of practical interest and have great potential for applications in microelectronics, magnetic memory, energy harvesters, and sensor technology [13, 48]. These materials are essentially magneto-electric transducers and can be used to generate and detect fields, especially *ac* and *dc* magnetic fields. Currently, Hall Probes, which work on the basis of giant magneto-resistance, are used to detect *dc* magnetic fields in spintronic devices. This measurement is based on transmission of current that leads to Ohmic losses. For the detection of *ac* magnetic fields, inductive sensors are used. But this measurement has a drawback of the dependence of electric response on frequency of the field. The problem of generation of magnetic field has been solved by using solenoid. But in this case, the inductive elements used in memory and other electronic devices hardly satisfy the ever-increasing miniaturization requirements. A natural solution to this problem is the use of linear magneto-electric effect which is reversible. Using the magneto-electric effect we can both detect and generate magnetic fields. Magneto-electric materials significantly reduce the losses of Joule heating as well as vertex current.

A magneto-electric effect is a tool for conversion of magnetic energy to electric energy and vice versa. The magneto-electric coupling factor is an important parameter for the indication of the usefulness of these materials for any device. Linear magneto-electric effect has a positive value, if the magnetic and electric fields are parallel or a negative value, if the fields are antiparallel, depending on the magneto-electric annealing in the case of polycrystalline materials. This can be used as the two states, 0, and 1, in the binary data storage devices as well as ROMs reading at high frequencies [7].

Multiferroics are also promising materials for spintronic devices. Spintronic devices are based on the transport properties of spin-polarized electrons. The existence of inhomogeneous magneto-electric interaction and micromagnetic in homogeneity in the material may generate non-uniform charge density distribution. This effect allows control of micromagnetic nanoaggregates by an electric field. This possibility is of undoubted interest for the development of both new media for magnetic data recording and spintronic devices [49].

Nonlinear M-E effect in composites also has a few applications. Since the field input to the composite is required to have two components, *dc* bias and *ac,* either of the two can be detected by providing the other component of a fixed magnitude in the input. Therefore, such magneto-electric composites can be used as magnetic probes for detecting *ac* as well as *dc* fields. So a magneto-electric composite can also compete in future with Hall probes in conventional applications of magnetic devices.

Magneto-electric materials should satisfy the following criteria for using them in micro-electronics and sensor technologies.

1. Temperatures of magnetic and electric orderings should exceed room temperature.
2. The value of Magneto-electriccoupling constant should be large (typically larger than 0.1 V/cm Oe).
3. The conductivity should be low.

The thickness of the face in layered composites can be controlled so that the resonant frequency is increased and they can be used in high frequency transducer applications. The hysteretic nature of the M-E composites may find application in Memory devices.

Magneto-electric nanomaterials (NMs) also have potential for applications in microelectronic industry. Moreover, magnetic NPs with functionalized surfaces have recently found many applications in biology, medicine, and biotechnology. Some such applications are *in vitro,* such as magnetic tweezers and magnetic separation of proteins and DNA molecules. Therapeutic applications include hyperthermia, targeted delivery of drugs and radioactive isotopes for chemotherapy and radiotherapy and contrast enhancement in magnetic resonance imaging [50].

11.15 FUTURE DIRECTIONS FOR 0-3 PARTICULATE MAGNETO-ELECTRIC SYSTEMS

Though magneto-electric effect in materials is proposed for many applications, no commercial applications have been realized yet for these composites. This may be due to the difficulty of achieving an ideal material obeying product property with reproducibility. The magneto-electric effect obtained so far in ceramic composites is ten times or more less than predicted theoretically. To achieve maximum product property, good mechanical coupling is required between two phases. Mechanical defects such as pores in the interface region between two phases should not exist in the composite for good mechanical coupling. At the same time, no chemical reaction should occur between the phases. Small deviations in composites while sintering may cause a reasonable drop in piezoelectric or magneto-electric properties of the composite phases. The connectivity pattern of the component phases in the composites is important for electric properties such as resistivity, dielectric constant, piezoelectric coefficient and hence the magneto-electric coefficient. In 0–3 magneto-electric composites good dispersion of the ferrite particles in the matrix is highly required in order to sustain sufficient electric resistivity of the composite. So in particulate composites prepared by the cost-effective route of high-temperature sintering, there is a challenge of obtaining good dispersion of ferrite in piezo matrix, favorable coherent interfaces, and sufficient bulk density by avoiding the reaction and interfacial diffusion between the two ceramic phases. Nowadays many research groups have adopted spark plasma sintering and hot pressing to avoid problems due to normal sintering.

Need of good alternatives to toxic lead-containing piezoceramics is also challenging to researchers in this field. Many researches are going on to replace lead with environment-friendly piezoceramic material with good piezoelectric and coupling properties.

Magneto-electric nanostructures or magneto-electric NCs have become an important topic of ever-increasing interest during the last few years. The potential of these materials and their applications are limitless. However, such studies are still in the infant stage. The first challenge in the field of magneto-electric NCs is the preparation of nanostructures persisting good electric and magnetic properties in nanoscale and having narrow size distribution to get high quality magneto-electric nanocomposite of piezo/ magnetic nanoparticle in magnetic/piezo matrix.

A great deal of theoretical work has already been devoted to the problem of understanding magneto-electric interactions in composites. Many simulations of these from micromechanics can be found in literature. Theoretical and experimental studies in NCs based on coupling between NPs and the matrix, effect of nanoscale coupling on material properties, critical dimension of NPs below which the properties change and coupling disappears, and dependence of parameters like electric and magnetic fields, magneto-electric, and magneto-capacitance effects, etc. are of great interest. Studies based on changes in the properties of materials on going from composites to NCs and the influence of interfacial effects on the properties of the material due to large interfacial area are also interesting and growing.

KEYWORDS

- **electronic industries**
- **ferroelectricity**
- **lead-free multiferroic magneto-electric composites**
- **lead-free piezoelectrics**
- **magneto-capacitance effects**
- **multiferroics**

REFERENCES

1. Fulde, P., (2002). *Electron Correlations in Molecules and Solids* (3rd edn.). Springer, Berlin.
2. Fazekas, P., (1999). *Lecture Notes on Electron Correlation and Magnetism.* World Scientific, Singapore.
3. Curie, P., (1894). On symmetry in physical phenomena, symmetry of an electric field and a magnetic field. *J. Phys., 3*, 393–415.
4. Hill, N. A., (2000). Why, are there so few magnetic ferroelectrics? *J. Phys. Chem. B., 104*(29), 6694–6709.
5. Pickett, W. E., & Spaldin, N. A., (2005). *Solid State and Material Science, 9*(3), 128–139.
6. Spaldin, N. A., & Pickett, W. E., (2003). Computational design of multifunctional materials. *J. Solid State Chem., 176*(2), 615–632.
7. Ryu, J., Priya, S., Uchiono, K., & Kim, H. E., (2002). Magneto-electric effect in composites of magnetostrictive and piezoelectric materials. *J. Electroceramics, 8*, 107–119.
8. Fiebig, M., (2005). Revival of the magneto-electric effect. *Journal of Physics D Applied Physics, 38*, R123–R152.
9. Velev, J. P., Jaswal, S. S., & Tsymbal, E. Y., (2011). Multi-ferroic and magneto-electric materials and interfaces. *Phil. Trans. R. Soc. A., 369*, 3069–3097.
10. Wang, Y., Hu, J., Lin, Y., & Nan, C. W., (2010). Multiferroic magneto-electric composite nanostructures. *NPG Asia Mater, 2*(2), 61–68.
11. Suchtelen, J. V., (1972). Product properties: A new application of composite materials. *Phil. Res. Rep.*, pp. 27–28.
12. Newnham, R. E., Skinner, D. P., & Cross, L. E., (1978). Connectivity and piezoelectric-pyroelectric composites. *Mater. Res. Bull., 13*, 525–536.
13. Nan, C. W., Bichurin, M. I., Dong, S., Viehland, D., & Srinivasan, G., (2008). Multiferroic magneto-electric composites: Historical perspective, status, and future directions. *J. Appl. Phys., 103*, 1–35.

14. Boomgaard, J. V. D., Van Run, A. M. J. G., & Suchetelene, J. V., (1976). Magneto-electricity in piezoelectric-magnetostrictive composites. *Ferroelectrics, 10,* 295–298.

15. Boomgaard, J. V. D., & Born, R. A. J., (1978). A sintered magneto-electric composite material $BaTiO_3$-Ni (Co. Mn) Fe_2O_4. *J. Mater. Sci., 13,* 1538–1548.

16. Boomgaard, J. V. D., Terrell, D. R., Born, R. A. J., & Giller, H. F. J. I., (1974). An *in situ* grown eutectic magneto-electriccomposite material. Part I: Composition and unidirectional solidification. *J. Mater. Sci., 9,* 1705–1709.

17. Ryu, J., Carazo, A. V., Uchino, K., & Kim, H. E., (2001). Piezoelectric and magneto-electric properties of lead zirconate titanate/Ni-ferrite particulate composites. *J. Electroceramics, 7,* 17–24.

18. Wang, K. F., Liu, J. M., & Ren, Z. F., (2009). Multiferroicity: The coupling between magnetic and polarization orders. *Advances in Physics, 58*(4), 321–448.

19. Eerenstein, W., Mathur, N. D., & Scott, J. F., (2006). Multiferroic and magneto-electric materials. *Nature, 442,* 759–765.

20. Zhai, J., Cai, N., Shi, Z., Lin, Y., & Nan, C. W., (2004). Magnetic-dielectric properties of $NiFe_2O_4$/PZT particulate composites. *J. Phys. D: Appl. Phys., 37,* 823–827.

21. Maaz, K., Karim, S., Mumtaz, A., Hasanain, S. K., Liu, J., & Duan, J. L., (2009). Synthesis and magnetic characterization of nickel ferrite nanoparticles prepared by co-precipitation route. *J. Magnetism and Mag. Mater., 321,* 1838–1842.

22. Guo, Y., Kakimoto, K., & Ohsato, H., (2004). Phase transitional behavior and piezoelectric properties of (Na0.5K0.5)NbO3-LiNbO3 ceramics. *Appl. Phys. Lett., 85,* 4121–4123.

23. Grossinger, R., Giap, V. D., & Sato-Turtelli, R., (2008). The physics of magneto-electric composites. *J. Magnetism and Magnetic Materials, 320,* 1972–1977.

24. Rakhikrishna, R., Isaac, J., & Philip, J., (2015). Magneto-electric characterization of $x(Na_{0.5}K_{0.5})_{0.94}Li_{0.06}NbO_3$-(1-x) $NiFe_2O_4$ composite ceramics. *J. Electroceram., 35,* 120–128.

25. Liu, W. C., Mak, C. L., Wong, K. H., Lo, C. Y., Or, S. W., Zhou, W., Hauser, A., Yang, F. Y., & Sooryakumar, R., (2008). Magneto-electric and dielectric relaxation properties of the high Curie temperature composite $Sr_{1.9}Ca_{0.1}NaNb_5O_{15}$–$CoFe_2O_4$. *J. Phys. D: Appl. Phys., 41,* 1–4.

26. Zhang, S., Li, L., & Kumar, A., (2008). *Materials Characterization Techniques* (pp. 125–131). CRC press, Boca Raton.

27. Jacob, B. P., Kumar, A., Pant, R. P., Singh, S., & Mohammed, E. M., (2011). Influence of preparation method on structural and magnetic properties of nickel ferrite nanoparticles. *Bull. Material Sci., 34,* 1345–1350.

28. Kambale, R. C., Song, K. M., & Hur, N., (2013). Dielectric and magneto-electric properties of $BaTiO_3eCoMn0.2Fe1.8O4$ particulate (0–3) multiferroic composites. *Curr. Appl. Phys., 13,* 562–566.

29. Fawzi, A. S., Sheikh, A. D., & Mathe, V. L., (2010). Multiferroic properties of Ni ferrite—PLZT composites. *Physica B., 405,* 340–344.

30. Sadhana, K., Murthy, S. R., Jie, S., Xie, Y., Liu, Y., & Li, Q. W., (2013). Magnetic field induced polarization and magneto-electric effect of Ba0.8Ca0.2TiO3-Ni0.2Cu 0.3Zn0.5Fe2O4 nanomultiferroic. *J. Appl. Phys., 113,* 1–3.

31. Bhadra, D., Masud, M. G., De, S. K., & Chaudhuri, B. K., (2012). Large magneto-electric effect and low-loss high relative permittivity in 0–3 CuO/PVDF composite

films exhibiting unusual ferromagnetism at room temperature. *J. Phys. D: Appl. Phys., 45*, 1–8.

32. Patankar, K. K., Dombale, P. D., Mathe, V. L., Patil, S. A., & Patil, R. N., (2001). AC conductivity and magneto-electric effect inMnFe1.8Cr0.2O4-BaTiO3 composites. *Mater. Sci. Eng. B., 87*, 53–58.

33. Sirdeshmukh, L., Kumar, K. K., Laxman, S. B., Krishna, A. R., & Sathaiah, G., (1998). Dielectric properties and electrical conduction in yttrium iron garnet (YIG). *Bull. Mater. Sci., 21*(3), 219–226.

34. Ramana, M. V., Reddy, N. R., Sreenivasulu, G., Kumar, K. V. S., Murty, B. S., & Murty, V. R. K., (2009). Enhanced mangnetoelectric voltage in multiferroic particulate $Ni_{0.83}Co_{0.15}Cu_{0.02}Fe_{1.9}O_4$–δ/PbZr$_{0.52}Ti_{0.48}O_3$ composites-dielectric, piezoelectric and magnetic properties. *Curr. Appl. Phys., 9*(5), 1134–1139.

35. Maxwell, J. C., (1973). *Electricity and Magnetism.* Oxford University Press, London.

36. Wagner, K. W., (1973). The distribution of relaxation times in typical dielectrics. *Ann. Physic, 40*, 817–819.

37. Koops, C. G., (1951). On the dispersion of resistivity and dielectric constant of some semiconductors at audio frequencies. *Phys. Rev., 83*, 121–124.

38. Kadam, S. L., Patankar, K. K., Mathe, V. L., Kothale, M. B., Kale, R. B., & Chougule, B. K., (2002). Dielectric behavior and magneto-electric affect in Ni0.75Co0.25Fe2O4 +Ba0.8Pb0.2TiO3 me composites. *J. Electroceram., 9*, 193–198.

39. Zhou, J. P., Lv, L., Liu, Q., Zhang, Y. X., & Liu, P., (2012). Hydrothermal synthesis and properties of NiFe2O4@BaTiO3 composites with well-matched interface. *Sci. Technol. Adv. Mater., 13*, 1–12.

40. Cimoga, C. E., Dumitru, I., Mitoseriu, L., Galassi, C., Iordan, A. R., Airimioaei, M., & Palamaru, M. N., (2010). Magneto-electric ceramic composites with double-resonant permittivity and permeability in GHz range: A route towards isotropic metamaterials. *Scr. Mater., 62*, 610–612.

41. Jigajeni, S. R., Tarale, A. N., Salinkhe, D. J., Kulkarni, S. B., & Joshi, P. B., (2012). Magneto-electric and magneto dielectric properties of SBN-CMFO nanocomposites. *Appl. Nanosci., 2*, 275–283.

42. Guo, Y., Liu, Y., Wang, J., Withers, R. L., Chen, H., Jin, L., & Smith, P., (2010). Giant magneto dielectric effect in 0-3 Ni0.5Zn0.5Fe2O4-Poly(vinylidene-fluoride) nanocomposite films. *J. Phys. Chem. C, 114*, 13861–13866.

43. Catalan, G., (2006). Magneto-capacitance without magneto-electric coupling. *Appl. Phys. Lett., 88*, 1–3.

44. Zhou, Y., Ye, Y. X., Zhou, S. H., Feng, Z. J., Yu, S. J., Chen, M. G., & Zhang, J. C., (2011). Frequency and field dependence of magneto-electric coupling in multiferroic particulate composites. *Eur. Phys. J.: Appl. Phys., 56*, 1–5.

45. Kulawik, J., Szwagierczak, D., & Guzdek, P., (2012). Magnetic, magneto-electric and dielectric behavior of CoFe$_2$O$_4$–Pb(Fe$_{1/2}$Nb$_{1/2}$)O$_3$ particulate and layered composites. *J. Mag. Mag. Mater., 324*, 3052–3057.

46. Rakhikrishna, R., Isaac, J., & Philip, J., (2017). Magneto-electric coupling in multi-ferroic nanocomposites of the type x(Na$_{0.5}$K$_{0.5}$)$_{0.94}$Li$_{0.06}$NbO$_3$- (1-x) CoFe$_2$O$_4$: Role of ferrite phase. *J. Ceram. Int., 43*, 664–671.

47. Guzdek, P., Sikora, M., Góra, L., & Kapusta, C. Z., (2012). Magnetic and magneto-electric properties of nickel ferrite-lead iron niobate relax or composites. *J. Eur. Ceram. Soc., 32,* 2007–2011.

48. Palneedi, H., Annapureddy, V., Priya, S., & Ryu, J., (2016). Status and perspectives of multiferroic magneto-electric composite materials and applications. *Actuators, 5*(9), 1–31.

49. Zvezdin, A. K., Logginov, A. S., Meshkov, G. A., & Pyatakov, A. P., (2007). Multiferroics: Promising materials for microelectronics, spintronics, and sensor technique. *Bull. Russ. Acad. Sci. Phys., 71,* 1561–1562.

50. Sukumaran, S., Neelakandan, M. S., Shaji, N., Prasad, P., & Yadunath, V. K., (2018). Magnetic nanoparticles: Synthesis and potential biological applications. *JSM Nanotechnol. Nanomed., 6*(2), 1068–1079.

CHAPTER 12

Behaviors and Green Properties of Interlocking Compressed Earth Bricks

ABDUL KARIM MIRASA, YVONNE WILLIAM TONDUBA,
R. EDDY SYAIZUL, AMIRA AMEER, HIDAYATI ASRAH, and
CHEE-SIANG CHONG

Faculty of Engineering, University Malaysia Sabah, Jln UMS–88400 Kota Kinabalu, Sabah, Malaysia, E-mails: akmirasa@ums.edu.my (A. K. Mirasa), drchongcs@gmail.com (C. S. Chong)

ABSTRACT

Fired clay brick (FCB), as a major building material in the construction industry, has been extensively produced and utilized around the world. Since the firing process for manufacturing FCB consumes a certain level of natural resources, the vast applications of FCB bring negative impacts to the environment. While the concern of awareness on the sustainability of building material and environmental pollution issues arises, the interlocking compressed earth brick (ICEB) has been innovated to replace FCB. Nevertheless, the investigation about the behavior and green properties of the ICEB is still limited. Much significant knowledge regarding ICEB production has not yet been explored. Thus, this research mainly aims to study the green properties of the existing ICEB and other potential green material for adding into ICEB to improve its green properties. This research also intends to investigate the behavior of the innovated ICEB. It can be concluded that the obtained research findings are encouraging. The acquired optimum mix design has attained maximum strength (about 5.70 N/mm^2), 17.5% water absorption and the soil is its major composition. These results are satisfactory in accordance with British Standard BS 3921:1985, and IS 1077–1992. Thusly, the acquired mix design of this study not only is able to attain the green properties, but also satisfactory for sustaining the structure loading.

12.1 INTRODUCTION

Fired clay bricks (FCB) are the widely used conventional bricks in Malaysia and worldwide. FCB is well known for its advantageous properties in terms of solid shape performance, higher fire resistance, attractive colors behavior, and low maintenance requirement. Since the fire process for producing the FCB not only requires the consumption of coal or other burning materials, but also releases several harmful gases to the air, the extensive production of FCB has bought many pollution issues to the environment. As the concern of awareness on the sustainability of building material and the environmental issues arises, many researchers have innovated the alternative material to improve the green properties of the conventional bricks.

Interlocking compressed earth brick (ICEB) is one of the green products that are highly potent to replace the role of FCB. While the production of ICEB requires only the high compression pressure and has avoided the fire process, it is environmental friendly, energy efficient and cost effective than the conventional FCB. The manufacture process of ICEB is considered as improved over traditional earth building techniques. In common, ICEB only consists of three main raw materials, which are sand, soil, and stabilizers. Water is added for activated the stabilizers to adhere the inherent soil particles to each other.

Similar to the manufacturing process of ICEB, the construction system using ICEB is also more effective and efficient in terms of cost and time aspect during the construction process if compare to those of FCB. As the continuity and strength of the ICEB have been improved by their interlocking mechanisms, the ICEB walling system has higher strength and is capable to take over the loading-bearing role. Thus, the construction system using ICEB can effectively eliminate the build-up of the conventional structural members, i.e., beam, and column. While the beams and columns are not necessary for sustaining the structural load, the ICEB construction system not only decreases the amount of reinforced concrete and but also eliminates all the slow and wasteful processes such as the installation of formwork.

Owing to the interlocking tongue and groove features of the ICEB system, the build-up process of the ICEB system is handy and no particular skill is required. It can be done in a dry stacking method. Thus, the overall cost of construction can be reduced by diminishing the usage of cement (where no plastering and mortar leveling needed for bricks stacking), the usage of reinforced concrete structural members and workers in constructing the building. In addition, faster completion can be accomplished with the ICEB system as compared to the conventional FCB system. All these beneficial behaviors are also contributed to the green properties to the ICEB system.

Furthermore, the main constituent of ICEB is soil. Soil, which is also known as earth material, is the oldest building material. Nowadays, earth material is seldom being used in the construction industry due to the development of several other modern building materials and methods. Soil can be termed as the most environmentally-friendly green building materials as it exists naturally almost everywhere and the extraction process of soil brings very low environmental impact. Besides, the soil is also durable, reusable, and recyclable.

Due to all these advantageous green behavior of ICEB, this research aims to investigate the behavior of the ICEB based on different mix design proportions. Particularly, the objective of this research can be broadly divided into two main concerns. The first objective intends to study the potential of green material (the natural soil) to be used in ICEB for improving its green properties. Secondly, investigations for ensuring the behavior of the ICEB in achieving the minimum requirement (in terms of physical and mechanical performances) are carried out.

In this chapter, the green properties of the ICEB are discussed in the first section. Subsequently, the raw materials and their particular behavior in affecting the performance of ICEB are briefly explained in Section 12.2. In order to increase the green properties of ICEB, numerous by-product of an industry that has the potential to produce the ICEB are described in Section 12.3. Subsequently, the manufacture process of ICEB is presented in Section 12.4. For clearly illustrating the methods to acquire the optimum mix design proportion of ICEB, the experiment set-up and tests are discussed in Section 12.5. Lastly, Section 12.6 presents the results and discussions of the tests and Section 12.7 concludes the research finding of the tests.

12.2   MATERIAL AND PROPERTIES

In general, the raw materials of ICEB are soil, sand, stabilizers, and water. The properties of the raw materials crucially influence the mechanical and durability properties of the produced ICEB. In the following sub-sections, the properties of these raw materials are briefly described.

12.2.1   SOIL AND SAND

Soil is a readily abundant natural resource in the world. It is the primary component in producing ICEB. The ultimate behavior of an ICEB is highly dependent on the properties of soil. Since soil is an earthen material, its properties vary based on location [41]. The important soil properties that affect the quality of ICEB are: soil particle distribution, clay content, and the plasticity index. These three soil properties are commonly used to classify the type of soil [40].

Particle size distribution of soil is the main parameter that influences ICEB properties. Soil as available in nature is quite variable in size distribution; it depends on the type of parent rock material. Generally, the available soil is high in clay content, which forms the finest soil particles. Therefore, it is rarely suitable to be directly used for production of ICEB as they do not meet the grading requirements for producing good quality brick [30]. Particle size analysis of soil is conducted by sieving method based on standard ASTM D422. Previous researcher suggested that, in order to meet the required particle size distribution, natural soil should be blended with some frictional material like sand which also acts as an aggregate. Well-mixed combination of clay (small aggregates) and sand (large aggregates) can be achieved when clay-filled up the empty space in between the sand particles [5]. Thus, sand is utilized as frictional material to improve the properties of the produced ICEB in this research.

Soil with higher clay content usually has higher soil plasticity index, therefore, sand is added to effectively reduce the plasticity index of soil. Plasticity index of soil is determined by the liquid limit and plastic limit of the soil. As the plasticity index of soil increases, the strength of ICEB significantly reduces [28]. According to Walker [46], the most ideal plasticity index of soil in fabricating ICEB is between 5 to 15. If the plasticity index of soil is larger than 20, it is not apt to be cement stabilized due to the problems of low compressive strength, insufficient durability and extensive drying shrinkage. In this research, the plasticity index of soil is examined based on standard ASTM D4318 and should has the acceptable plasticity index. Figure 12.1 shows the site where the soil has been excavated to produce the ICEB and the natural drying process in this research.

FIGURE 12.1   Site for extracting soil and the drying process for soil of this research.

12.2.2   STABILIZER

Stabilizer is added into the mixture of ICEB to improve the characteristic of the produced brick. It plays an important role in creating bonding between soil-stabilizers mixes [35]. There are two types of brick stabilizers, which are ordinary Portland cement (OPC) and lime. The most common stabilizer used for manufacturing the ICEB is OPC. Cement is preferred for low plasticity (sandy) soil but it is not suitable for highly plastic soil. Therefore, the combination of clay and soil is significant for brick mixture design. The recommended cement content to be added in the mixture is within of 5 to 12% of soil mass, as higher amount of cement content also increases the costing of the brick production [45]. With adequate stabilizer content, it is able to ensure adequate strength, lower erosion rate and lower water absorption of ICEB [27]. Therefore, OPC content utilized in this research is maintained to satisfy the behavior and cost requirement of the bricks.

12.2.3   WATER

Water is needed in the mix design to activate the hydration process of OPC. Moreover, clay that presents in the soil also requires moisture to reach its plasticity state, or the state in which it binds together and becomes moldable [41]. However, optimum water content of the mixture needs to be investigated. If the water content of the mixture is too low, there is insufficient moisture for the mixture to be reactive and thus, the strength of ICEB is affected. Whereas the excessive water content affects the molding of ICEB, as the soil sticks on the wall of the mold. The optimum water content is needed to mold the ICEB and eject them successively as one unit [31]. In this research, ASTM D558 is used to determine the optimum water content for soil-cement specimens.

12.3   OTHER POTENTIAL GREEN MATERIAL IN PRODUCING INTERLOCKING COMPRESSED EARTH BRICKS (ICEBS)

12.3.1   PALM OIL FUEL ASH (POFA)

Palm oil fuel ash (POFA) is a waste product from the palm oil mill. It is a product from the incineration of palm fibers and palm shells to produce steam for electrical generation. Figure 12.2 shows POFA that has been sieved through 150 μm. POFA contains high amounts of silicon and aluminum oxides in the amorphous state and has been accepted as a pozzolanic material [42]. Table 12.1 shows the physical and chemical properties of POFA based on previous research studies. The utilization of POFA is minimal compared to the amount produced. Most POFA are disposed in landfills, resulting in environmental pollution and other problems. This industrial waste disposal has always been a concern in Malaysia. The total solid waste generated by the industry in Malaysia amounts to about 10 million tonnes a year [36].

There are few research studied on usage of POFA in production of brick. Ismail et al. [23] did not use soil to produce their brick; rather they fabricated the bricks from cement, sludge, and POFA [23]. Nasly et al. [32] uses POFA as partial sand replacement in the production of interlocking block work system [32]. Rahman et al. [34] uses POFA as a partial replacement of cement in production of masonry block [34]. Kadir et al. [24] incorporated POFA in FCBs [24].

FIGURE 12.2 Palm oil fuel ash (POFA).

TABLE 12.1 Physical and Chemical Properties of POFA

Researcher(s)	Alsubari et al. [7]	Hussin et al. [20]	Awal et al. [8]	Noorvand et al. [33]	Kakhuntodd et al. [26]
Specific Gravity	2.44	2.18	2.42	1.97	2.43
Blaine Fineness (cm^2/g)	–	–	4930	–	–
Specific Surface Area (BET), (m^2/g)	–	14.4	–	25.5	–
Median Particle size, d$_{50}$ (µm)	–	–	–	82	8.1
Retained on 45 µm (%)	4	1	10.5	45	1
Silicon dioxide (SiO$_2$)	59.17	53.82	59.62	48.9	57.8
Aluminum oxide (Al$_2$O$_3$)	3.73	5.66	2.54	2.71	4.6
Iron oxide (Fe$_2$O$_3$)	6.33	4.54	5.02	6.54	3.3
Calcium oxide (CaO)	5.8	4.24	4.92	13.89	6.6
Magnesium oxide (MgO)	4.87	3.19	4.52	2.74	4.2
Sulfur trioxide (SO$_3$)	0.72	2.25	1.28	1.54	0.3
Potassium oxide (K$_2$O)	8.25	4.47	7.52	7.13	8.3
Sodium oxide (Na$_2$O)	0.18	0.1	0.76	0.05	0.5
Phosphorus pentoxide (P$_2$O$_5$)	–	3.01	1.28	3.67	–
Loss on ignition (LOI)	16.1	10.49	8.25	11.3	10.1
SiO$_2$+Al$_2$O$_3$+Fe$_2$O$_3$	69.23	64.02	67.18	58.15	65.7

Based on their results, POFA have shown significant influence towards the properties of brick. In term of physical and mechanical properties, the compressive strength decreased with increasing of POFA content [34]. The incorporation of palm oil waste required higher water content to maintain the flowability of the brick mixture [24]. Apart from that, the increment in POFA replacement causing the surface of brick more brittle [32]. Rahman explained that higher POFA replacement was related to a weaker bonding between the POFA, cement, and aggregate. While the impact towards the durability properties, ICEB that are fabricated from cement, POFA, and paper sludge have strong chemical stabilization abilities in suppressing the release of unacceptable amount of heavy metal through leaching [23]. However, the water absorption of POFA masonry blocks is observed to be slightly higher (0.1%) [34]. The reason behind this is due to the greater porosity of POFA which tends to favor water absorption [36]. Rahman investigated POFA was able to prevent the efflorescence due to its pozzolanic properties. Efflorescence is the formation of salt deposits on the masonry block surfaces due to leaching of lime compounds.

12.3.2 QUARRY DUST

Quarry dust is a by-product from the aggregate crushing process, which forms in rubble crusher units [11, 12, 18, 25, 37]. During the blasting activities, rocks are crushed into various sizes. Dust that is generated from the process is known as quarry dust and shown in Figure 12.3. Quarry dust is the greyish fine rock particles with size below 5 mm [19]. Quarry dust caused the landfill disposal problems, health, and environmental pollution [25]. From the literature, the physical and chemical properties of quarry dust are shown in Tables 12.2 and 12.3.

Based on previous studies, quarry dust is commonly used as an alternative material for sand replacement in concrete production due to its physical and chemical properties that have satisfied the requirements of fine aggregate. Quarry dust shows more silicates in chemical composition compared to other compounds. Thus, during the hydration process, it is expected that the silicates improve the bonding process. Compared to natural sand, quarry dust has alike properties. The range of specific gravity for quarry dust is within 2.3–3.0 to be used as a sand replacement. The quarry dust shall be passing by No. 200 sieve for cement replacement and retaining by No. 100 sieve as a sand replacement [29].

FIGURE 12.3   Quarry dust.

TABLE 12.2   Physical Properties Comparison of Coarse Aggregates, Mining Sand, and Quarry Dust

Properties	Coarse Aggregate	Mining Sand	Quarry Dust
Maximum size (mm)	20.0	5.00	5.00
Specific gravity	2.65	2.73	2.83
Absorption (%)	0.55	0.74	0.78
Moisture content (%)	0.81	0.97	0.94
Fineness modulus	7.26	3.24	3.41
Silt content (%)	-	5.7	13.0

Source: Raman et al. [48].

The strength of concrete increases with the replacement of quarry dust as sand if compare to the concrete made with equal quantities of river sand [19]. Same with ordinary concrete, the compressive strength of quarry dust concrete continues with age. The use of quarry dust in concrete can improve the cost effective of concrete as it reduces the strain on supply of natural fine aggregate. Yet, the use of quarry dust in mixes deflects the workability. This limitation can be overcome by adding the fly ash or chemical admixtures [19]. Dehwah [12] recommends that the utilization of quarry dust in regions where sand is not easily available. Replacement of sand by the quarry dust can be up to 40% in enhancing the compressive strength. With increasing the replacement percentage, the workability of the concrete decreases [38]. In contrast to FCB, the best percentage of quarry dust is 10% for obtained optimum physical and mechanical properties of the clay brick [25].

TABLE 12.3   Chemical Properties of Quarry Dust

Researcher(s)	Shakir et al. [37]	Thomas and Harilal [43]	Hamid Mir [19]
Silicon dioxide (SiO_2)	69.94	62.5	62.48
Aluminum oxide (Al_2O_3)	14.60	18.7	18.72
Iron oxide (Fe_2O_3)	2.16	6.5	6.54
Calcium oxide (CaO)	2.23	4.8	4.83
Magnesium oxide (MgO)	0.38	2.5	2.56
Sulfur trioxide (SO_3)	–	1.0	–
Potassium oxide (K_2O)	–	–	3.18
Loss on ignition (LOI)	0.74	–	0.48
$SiO_2+Al_2O_3+Fe_2O_3$	86.70	64.02	67.18

12.3.3   ECO PROCESSED POZZOLAN (EPP)

Pozzolanic materials are commonly used for partially replacing OPC to impart properties of concrete or mortar depending on the demand of a user, and to reduce the cost due to the reduction in the use of OPC in the concrete mixture [6]. Besides POFA, rice husk ash, fly ash, and any other pozzolanic materials that were studied, EPP is one of alternative materials that can be used as an addition in ICEB mixture to replace the percentages of cement in order to produce more economical as well as ecological ICEB.

During the degumming and bleaching process of crude palm oil (CPO), bleaching earth is used to remove the impurities [13] such as color, pigments, and other impurities that harmful to people's health [44]. This process produces a refined, bleached, and deodorized (RBD) palm oil as well as a spent bleaching earth (SBE) which is classified as solid waste and causes a problem as it is a waste material and normally thrown into landfills. Through a process of extracting and refining SBE, the SBE will absorbs some of the oil and when the oil is recovered SBE will produced another product such as spent bleaching oil (SBEO), Eco-Mineral, and a sustainable product called EPP. EPP mainly used in construction industry as a replacement materials for cement [13]. Figure 12.4 shows the SBE converted to another product which are SBEO, eco-mineral, and EPP.

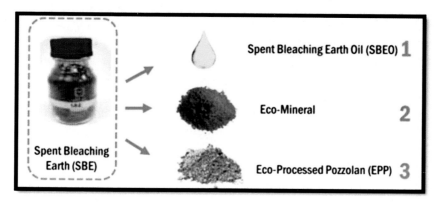

FIGURE 12.4 Spent bleaching earth (SBE) converted to another products which are SBEO, eco-mineral, and EPP.

Source: EcOoils Sdn. Bhd [13].

EPP falls into the category of "calcined natural pozzolan for use in concrete" with standard ASTM C618 –08 A (Class N) where the EPP acts as a great binding agent to improve the strength, compactness, chemical resistance, and durability of mixture. According to the finding of EcoOils, EPP also improves coherence, stability, and compactness as a filling agents as well as water permeability coefficient and affinity to Cationic Metals as an adsorbent agent. Tables 12.4 and 12.5 show the physical and chemical characteristics of EPP respectively.

According to Yetgin and Çavdar [47], pozzolans can be added to cement during the production process as well as can be directly mixed into concrete. Besides that, 30% of EPP as a cement replacement in foamed concrete gave out a better performance properties compared to control sample [49]. Thus, it is predictable that EPP can increase the strength of ICEB with an optimum design.

12.4 GEOMETRY AND MANUFACTURING FOR ICEB

In overall, four major processes, which include of crushing, mixing, compacting, and curing, are involved in the factory process for producing the ICEB. At first, the extracted soil (from the site) must be dried for 24 to 48 hours before proceeding to the crushing process. If the soil is not adequately dried, it would stick and jammed the crusher. After the soil is crushed, it is delivered to the mixer drum through the conveyor. Subsequently, all the raw materials are poured into the mixer drum and mixed in dried stage. This is to ensure that the materials are homogenously mixed. Then water is added accordingly to the mixture as the wet mixing process. While the mixture is mixed thoroughly, they are delivered to the compacting machine where high compaction pressure is applied. During compaction, the pressure must be consistent; else, the dimension of the bricks is affected. Lastly, the produced ICEB are stacked for curing process. The ICEB are allowed to be dried for 24 hours before curing process starts. The ICEB are cured by using the water spray so that the surface of the bricks not eroded by the water pressure. Curing is continuously done with minimum of14 days.

TABLE 12.4   Physical Properties of EPP

Parameter	Result
Appearance	Free Flowing
+45 microns,%	20–24 (34 Max)
Bulk Density, mt/m³	0.58
Density, mt/m³	2.0–2.1
Reactive Silica,%	25 Min
Ash,%	98 Min

Source: EcOoils Sdn. Bhd [13].

TABLE 12.5   Chemical Properties of EPP

Parameter	Benchmark	Results
$SiO_2+Al_2O_3+Fe_2O_3$	% Min, 70	75–85
SO_3	% Max, 4	1.0–1.5
Moisture	% Max, 3	0.5 Max
LOI	% Max, 10	2.0 Max

Source: EcOoils Sdn. Bhd [13].

Figure 12.5 has illustrated the dimensions and types of the ICEB which is produced in this research. All types of ICEB have hollow center which intends to reduce the weight of the brick, insert steel rod or bar for reinforcement, act as conduit for electrical and water piping and for grouting which can increase the stability and providing barrier to seepages. In this research, ICEB can be compressed with three types of shape, i.e., standard brick, half brick, and U-brick. Each shape has its individual function in constructing the building. Standard bricks are major use in main structural element such as wall or column. Different from U-brick which act as a beam and used as a horizontal ring beam in every five-layer bricks in wall. For half brick, most of the function is to fill every corner of the wall and column element.

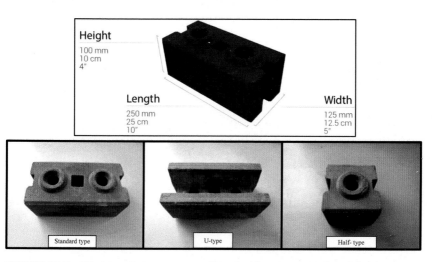

FIGURE 12.5 The manufacture process of interlocking compressed earth bricks.

12.5 EXPERIMENTAL SET-UP AND TESTS

Principally, the physical and mechanical properties of bricks are the vital parameters to determine its performances in constructing the building. The major mechanical properties that play the main roles in affecting the behavior of bricks are compressive strength and water absorption properties. The experimental set-up and tests for examining these aforementioned properties are discussed in the following sub-sections.

12.5.1 COMPRESSION TEST

According to Ben Ayed et al. [9], many researchers had proposed various testing procedures to conduct the compression test on bricks and blocks, for instance, RILEM test, half block test, cylindrical test, and friction test. While the standard procedures for testing compressive strength are also being stated clearly in Eurocode, EN 772-1:2000, EN 1052-1:2009 and British code, BS 3921-1985, there are many constructive guidelines for carrying out the compression test. Basically, the brick samples of this study were prepared and tested based on those mentioned in EN 772-1:2000 for evaluating their compressive strength.

As mentioned in clause 7.1 of EN 772-1:2000, the minimum number of brick specimens shall be six. Thus, at least six ICEB samples were prepared for evaluating the compressive strength of the ICEB. EN 772-1:2000 also states that the surface of the samples can be prepared by either grinding (clause 7.2.4) or capping (clause 7.2.5). The ICEB samples of this research were tested by the capping method. Table 12.2 (in Section 12.2 of EN 772-1:2000) also provides the loading rate for testing the samples. As the expected compressive strength of the samples are less than 10 N/mm^2, the loading rate shall be about 0.05 $(N/mm^2)/s$. Figure 12.6 illustrates the laboratory set-up of the compression test. The results of the testing are described thoroughly in Section 12.6.

FIGURE 12.6   The experimental set-up of compression test.

12.5.2   *WATER ABSORPTION TEST*

Water absorption property of a masonry unit is another essential parameter that influences its performances in constructing a building. As the water absorption behavior is factor that major impairs the quality as caused the problem like surface erosion, peeling off of surface finishes and even partial crumbling in certain severe cases [17], the water absorption properties of samples should be examined. Similar to compression test, there are several standard tests can be followed for determining the water absorption of brick sample, i.e., Euro code EN 772-7, British Code BS 3921-1985, Indian Standard IS 3495-2:1992.

Although BS 3921-1985 provides guidelines for testing the water absorption using the 5 hours boiling test, it has included the 24-hour cold immersion and absorption as the control test. The water absorption of the samples in this research was determined based on the Indian Standard IS 3495-2:1992, which tests the ICEB samples with 24-hour cold immersion and absorption. Figure 12.7 has shown the oven drying process of the ICEB, the 24-hour cold water immersion and weight measurement of the ICEB samples. The procedures of the testing are followed mainly as those stated in IS 3495–1985, whereas the acquired results are described in Section 12.6.

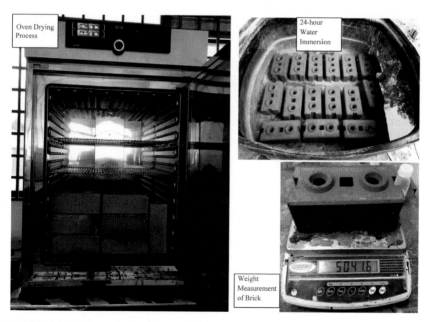

FIGURE 12.7   The experimental set-up of the water absorption test.

12.6   RESULTS AND DISCUSSION

As aforementioned in Section 12.5, the physical and mechanical properties of bricks crucially affect its behavior for construction purposes. Hence, geometry and density measurements of the ICEB samples are significant to be discussed before other properties. In order to clearly present the geometry properties of the samples, Figure 12.8 has been provided for demonstrating the physical performances of the ICEB samples in this study. It is noticeable that the color of the ICEB samples generally turns from dark-grey to yellowish-brown as the percentage of the soil increases. However, the rectangle shape and the edges of the ICEB are considered well-maintained in all the cases.

FIGURE 12.8   The brick samples with various amounts of sand.

For bricks fabricated in hydraulic pressure compressed machine with a fixed dimension of mold that has constant volume, the density of the bricks may vary owing to the incorrect mixture content and varying moisture contents of soil. The incorrect mixture content and different moisture contents of soil may severely affect the strength of the produced brick samples [39]. Simion [39] also states that the constant volume method, which recommends to check the weight of the brick, is the most appropriate first test to ensure that the densities of the brick samples are within the satisfactory range for achieving the adequate compressive strength. Thus, weights of the ICEB samples in this research were measured on 7th, 14th, and 28th day and the respective results are listed in Table 12.6. It can be clearly observed that the variations of the weight are within 0.5 kg (10%) from 5 kg. When the mixture content of cement, sand, and soil are different, the weight differences of the produced ICEB samples are larger. This is acceptable as the density of the raw materials is diverse.

TABLE 12.6   Weight Measurements of the Brick Samples

Ratio (Cement:Sand:Soil)	% Soil	Weight (kg)		
		7th Day	14th Day	28th Day
1:0.60:4.33	73	4.98	5.08	5.21
1:1.19:3.74	63	4.87	4.85	4.91
1:1.78:3.14	53	4.95	4.92	5.08
1:2.38:2.55	43	4.59	4.62	4.66
1:2.97:1.95	33	4.54	4.53	4.66
1:3.57:1.36	23	5.18	5.13	5.16
1:4.16:0.76	13	5.12	5.05	5.12

As mentioned previously, the compressive strength of bricks plays an essential role to indicate its performance in constructing a structure, hence, the compressive strength of ICEB samples at 14th day and 28th day were tested and the results are tabulated in Tables 12.7 and 12.8 correspondingly. Due to the varied mixture content of soil and sand percentage, the acquired maximum load of the ICEB samples varied, respectively.

TABLE 12.7   Compressive Strength of the ICEB Samples at 14th day

Ratio (Cement: Sand: Soil)	Soil (%)	Average Max Load (KN)	Compressive Strength (N/mm$^{2)}$)
1:0.60:4.33	73	94.054	3.01
1:1.19:3.74	63	117.144	3.75
1:1.99:2.94	50	159.427	5.10
1:2.38:2.55	43	78.193	2.50
1:2.97:1.95	33	82.094	2.63
1:3.57:1.36	23	161.714	5.17
1:4.16:0.76	13	163.988	5.25

Table 12.8. Compressive Strength of the ICEB Samples at 28th Day

Ratio (Cement: Sand: Soil)	Soil (%)	Average Max Load (KN)	Compressive Strength (N/mm²)
1:0.60:4.33	73	127.091	4.07
1:1.19:3.74	63	128.906	4.12
1:1.99:2.94	50	178.073	5.70
1:2.38:2.55	43	112.725	3.61
1:2.97:1.95	33	100.509	3.22
1:3.57:1.36	23	243.067	7.78
1:4.16:0.76	13	188.191	6.02

In order to fulfill the requirement of EN 772-1:2000, 12 of ICEB specimens for each of different mixture content proportions were cast for carrying out the compression test on the 14th day and 28th day. In this study, the behavior of ICEB with 13% to 73% of soil were investigated, while the cement contents of these samples remained the same. For clearly illustrating the trend of the compressive strength for ICEB with various soil and sand proportion, Figure 12.9 is presented.

Primarily, Tables 12.7 and 12.8, Figure 12.9 has shown that the ICEB samples are capable to attain at least 80% strength of those on 28th day (except the ICEB samples with 60% sand, which attained about 67% of the strength on 28th day). From Figure 12.10, it also can be observed that the trend of the compressive strength can be broadly classified into three stages. When the soil percentage decrease from 73% to 50%, the compressive strength of the ICEB increases respectively. As discussed previously in Section 2.1, sand, which is utilized as frictional material, is competent to decrease the plasticity index and hence, increases the strength of the ICEB sample.

Yet, the compressive strength of ICEB decreases if the soil percentage decreases from 50% to 33%. Owing to the cohesiveness of the soil, the reduction of soil (when the soil content is a majority in the mixture) diminishes the adhesive properties of ICEB. Therefore, the decreasing of soil in this range decreases the compressive strength of the ICEB sample. However, if sand is the majority content in the mixture (more than 50%), the brittleness of ICEB due to the reduction of clay content (soil) decreases, and thus, the compressive strength behaves different than those of other two stages. Contrary to ICEB, if the percentage of soil is less than 33%, the sample is more like a mortar sample.

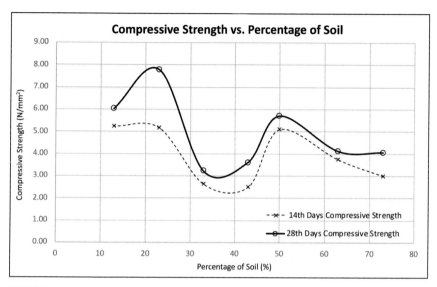

FIGURE 12.9   The compressive strength of ICEB with various amounts of sand.

Hence, the third batch of ICEB samples, which comprised of 33.6% sand, 50% soil, 16.4% cement have attained maximum strength (about 5.70 N/mm^2 on 28th day) and with the higher green properties as the soil are still the major composition of ICEB. In accordance with British Standard BS 3921:1985, the minimum compressive strength for the bricks with damp-proof course features has been specified as 5 N/mm^2. Moreover, IS 1077–1992 (clause 4.1) also restricts the minimum compressive strength of common brick as 3.5 N/mm^2. Thusly, the acquired mix design of this study is not only able to attain the green properties, but also satisfactory for sustaining the structure loading.

Successively, the water absorption results of the ICEB are discussed. The acquired water absorptions of the ICEB samples with respect to a different amount of sand percentage in this research study are presented in Figure 12.10. Overall, water absorption increases as the percentage of sand become higher. Owing to the larger porosity of sand if compare to those of clayey soil, the water absorption of ICEB samples with a higher amount of sand becomes larger.

As stated in Clause 7.2 of IS 1077–1992, when the bricks are tested based on the procedures specified in IS 3495-2: 1992 (where the bricks are immersed in cold water for 24 hours), the water absorption shall not be more than 20% by weight for common bricks. In reference to Figure 12.10, the water absorption of all the ICEB samples in this research is basically lesser than 20%. Therefore, the water absorption behavior of the produced ICEB samples has fulfilled the requirement as a common brick.

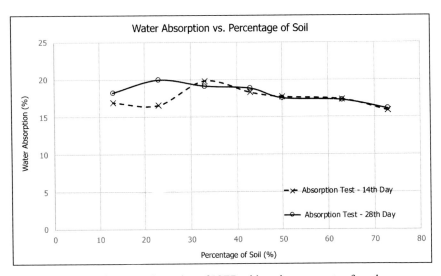

FIGURE 12.10   The water absorption of ICEB with various amounts of sand.

12.7   CONCLUSIONS

From this research, the green properties and behavior of the ICEB were investigated and encouraging findings were acquired.

- The potential of green materials (the natural soil) in producing the ICEB for improving its green properties has been validated.

  Due to the manufacturing and construction methods of the ICEB system, the green properties of ICEB are explored. For improving the structural performance of the ICEB system, the raw materials of ICEB and their properties are investigated. In order to further improve the green properties of ICEB, the material properties of other potential green material i.e., POFA, Quarry dust, and Eco Processed Pozzolan (EPP) are reviewed.

- The investigations for ensuring the behavior of ICEB in achieving the minimum requirement (in terms of physical and mechanical performances) have been conducted.

    For achieving the satisfactory performance, the mix designs of ICEB samples have been prepared with different proportion of sand and soil and their behaviors have been examined physically and mechanically. The physical behaviors (weight and solid shape) of the ICEB sample have the promising performances. The variation of weight for the cast samples are less than 10% from 5 kg. Based on the measured compressive strength of ICEB samples, the optimum design can be evaluated. Since 33.6% sand, 49.6% soil and 16.9% cement had accomplished the highest maximum compressive strength (about 5.7 N/mm^2 on 28th day) and contained the highest amount of green material (natural soil), it can be termed as the optimum design. The water absorption of the obtained optimum design samples, which is about 17.5%, also satisfied the common brick requirement of IS 1077–1992.

ACKNOWLEDGMENTS

The research team would like to express the deepest gratitude to the financial support from the Ministry of Higher Education (KPT) Malaysia under Translational Research Program, TR@M, Grant no. LRGS0008-2017. Sincere appreciation is also extended to all the members of the Faculty of Engineering, University Malaysia Sabah who had contributed their efforts to the research project.

KEYWORDS

- **earth break**
- **environmental pollutions**
- **green materials**
- **green properties**
- **interlocking compressed earth brick**
- **palm oil fuel ash**

REFERENCES

1. ASTM D422, (2007). *Standard Test Methods for Particle-Size Analysis of Soils.*
2. ASTM D4318, (2005). *Standard Test Methods for Liquid Limit, Plastic Limit, and Plasticity Index of Soils.*
3. ASTM D558, (2003). *Standard Test Methods for Moisture-Density (Unit Weight) Relations of Soil-Cement Mixtures.*
4. ASTM C6180-08. *Standard Specification for Coal Fly Ash and Raw or Calcined Natural Pozzolan for Use in Concrete.*
5. Abdullah, A. H., Nagapan, S., Antonyova, A., Rasiah, K., Yunus, R., & Sohu, S., (2017). Comparison of strength between laterite soil and clay compressed stabilized earth bricks (CSEBs). *MATEC Web of Conferences, 103.* https://doi.org/10.1051/matecconf/201710301029 (accessed on 22 February 2020).
6. Abdulmatin, A., Tangchirapat, W., & Jaturapitakkul, C., (2018). An investigation of bottom ash as a pozzolanic material. *Construction and Building Materials, 186,* 155–162. https://doi.org/10.1016/j.conbuildmat.2018.07.101 (accessed on 22 February 2020).
7. Alsubari, B., Shafigh, P., Jumaat, M. Z., & Alengaram, U. J., (2014). Palm oil fuel ash as a partial cement replacement for producing durable self-consolidating high-strength concrete. *Arabian Journal for Science and Engineering, 39*(12), 8507–8516. https://doi.org/10.1007/s13369-014-1381-3 (accessed on 22 February 2020).
8. Awal, A. S. M. A., & Shehu, I. A., (2013). Evaluation of heat of hydration of concrete containing high volume palm oil fuel ash. *Fuel, 105,* 728–731. https://doi.org/10.1016/j.fuel.2012.10.020 (accessed on 22 February 2020).
9. Ben, A. H., Limam, O., Aidi, M., & Jelidi, A., (2016). Experimental and numerical study of interlocking stabilized earth blocks mechanical behavior. *Journal of Building Engineering, 7,* 207–216.
10. British Standard Institution, British Standard Specifications for Clays Bricks, London, BS 3921, 1985.
11. Dayananda, N., & Keerthi, G. B. S., (2017). Prediction of properties of fly ash and cement mixed GBFS compressed bricks. *Materials Today: Proceedings, 4*(8), 7573–7578. https://doi.org/10.1016/j.matpr.2017.07.089 (accessed on 22 February 2020).
12. Dehwah, H. A. F., (2012). Mechanical properties of self-compacting concrete incorporating quarry dust powder, silica fume, or fly ash. *Construction and Building Materials, 26*(1), 547–551. https://doi.org/10.1016/j.conbuildmat.2011.06.056 (accessed on 22 February 2020).
13. EcoOils (2019). Product: Eco-Processed Pozzolan (EPP). http://www.ecooils-kcg.com/products (accessed 1 March 2020).
14. Safiuddin, M., Salam, M. A., & Jumaat, M. Z., (2011). Utilization of palm oil fuel ash in concrete: A review. *Journal of Civil Engineering and Management, 17,* 234–247. https://doi.org/10.3846/13923730.2011.574450.
15. Yetgin, S., & Cavdar, A., (2006). Study of effects of natural pozzolan on properties of cement mortars. *Journal of Materials in Civil Engineering, 18,* 813-816. https://doi.org/10.1061/(ASCE)0899-1561(2006)18:6(813).
16. EN 1052–1, (2009). *Methods of Test for Masonry.* Part 1: Determination of compressive strength.

17. Fadele, O. A., & Ata, O., (2018). Case study water absorption properties of sawdust lignin stabilized compressed laterite bricks. *Case Studies in Construction Materials.*
18. Galetakis, M., Piperidi, C., Vasiliou, A., Alevizos, G., Steiakakis, E., Komnitsas, K., & Soultana, A., (2016). Experimental investigation of the utilization of quarry dust for the production of micro cement-based building elements by self-flowing molding casting. *Construction and Building Materials, 107*, 247–254. https://doi.org/10.1016/j.conbuildmat.2016.01.014 (accessed on 22 February 2020).
19. Hamid, M. A., (2015). Improved concrete properties using quarry dust as replacement for natural sand. *International Journal of Engineering Research and Development, 11*(3), 2278–67.
20. Hussin, M. W., Muthusamy, K., & Zakaria, F., (2010). Effect of mixing constituent toward engineering properties of POFA cement-based aerated concrete. *Journal of Materials in Civil Engineering, 22*(4), 287–295. https://doi.org/10.1061/(ASCE)0899-1561(2010)22:4(287) (accessed on 22 February 2020).
21. IS 1077–1992. *Indian Standard Common Burnt Clay Building Bricks-Specification.*
22. IS 3494–2 (1992). *Indian Standard Methods of Tests of Burnt Clay Building Bricks.* Part 2: Determination of water absorption.
23. Ismail, M., Ismail, M. A., Lau, S. K., Muhammad, B., & Majid, Z., (2010). Fabrication of bricks from paper sludge and palm oil fuel ash. *Concrete, 1*, 60–66.
24. Kadir, A. A., Hassan, M. I. H., Sarani, N. A., Rahim, A. S. A., & Ismail, N., (2017). *Physical and Mechanical Properties of Quarry Dust Waste Incorporated Into Fired Clay Brick, 020040*, 020040. https://doi.org/10.1063/1.4981862 (accessed on 22 February 2020).
25. Kadir, A. A., Sarani, N. A., Abdullah, M. M. A. B., Perju, M. C., & Sandu, A. V., (2017). Study on fired clay bricks by replacing clay with palm oil waste: Effects on physical and mechanical properties. *IOP Conference Series: Materials Science and Engineering, 209*(1). https://doi.org/10.1088/1757-899X/209/1/012037 (accessed on 22 February 2020).
26. Kakhuntodd, N., Chindaprasirt, P., Jaturapitakkul, C., & Homwuttiwong, S., (2012). *The Investigation of Water Permeability of High Volume Pozzolan Concrete* (pp. 1–8).
27. Kariyawasam, K. K. G. K. D., & Jayasinghe, C., (2016). Cement stabilized rammed earth as a sustainable construction material. *Construction and Building Materials, 105*, 519–527. https://doi.org/10.1016/j.conbuildmat.2015.12.189 (accessed on 22 February 2020).
28. Kwon, H. M., Le, A. T., & Nguyen, N. T., (2010). Influence of soil grading on properties of compressed cement-soil. *KSCE Journal of Civil Engineering, 6*, 845–853. https://doi.org/10.1007/s12205-010-0648-9 (accessed on 22 February 2020).
29. Muhit, I. B., Raihan, M. T., & Nuruzzaman, M., (2015). Determination of mortar strength using stone dust as a partially replaced material for cement and sand. *Advances in Concrete Construction, 2*(4), 249–259. https://doi.org/10.12989/acc.2014.2.4.249 (accessed on 22 February 2020).
30. Nagaraj, H. B., Rajesh, A., & Sravan, M. V., (2016). Influence of soil gradation, proportion, and combination of admixtures on the properties and durability of CSEBs. *Construction and Building Materials, 110*, 135–144. https://doi.org/10.1016/j.conbuildmat.2016.02.023 (accessed on 22 February 2020).

31. Nagaraj, H. B., Sravan, M. V., Arun, T. G., & Jagadish, K. S., (2014). Role of lime with cement in long-term strength of compressed stabilized earth blocks. *International Journal of Sustainable Built Environment, 3*(1), 54–61. https://doi.org/10.1016/j.ijsbe.2014.03.001 (accessed on 22 February 2020).

32. Nasly, M. A., Yassin, A. A., & Nordin, N., (2011). Sustainable housing using an innovative mortar less interlocking blocks work system–the effect of palm oil fly ash (Pofa) as an aggregate replacement. *Proceedings of International Building and Infrastructure Technology Conference* (pp. 73–81). Retrieved from: http://www.hbp.usm.my/bitech2011/bitech2011proc.pdf (accessed on 22 February 2020).

33. Noorvand, H., Ali, A. A. A., Demirboga, R., Noorvand, H., & Farzadnia, N., (2013). Physical and chemical characteristics of unground palm oil fuel ash cement mortars with nanosilica. *Construction and Building Materials, 48,* 1104–1113. https://doi.org/10.1016/j.conbuildmat.2013.07.070 (accessed on 22 February 2020).

34. Rahman, M. E., Boon, A. L., Muntohar, A. S., Hashem, T. M. N., & Pakrashi, V., (2014). Performance of masonry blocks incorporating palm oil fuel ash. *Journal of Cleaner Production, 78,* 195–201. https://doi.org/10.1016/j.jclepro.2014.04.067 (accessed on 22 February 2020).

35. Riza, F. V., Rahman, I. A., Mujahid, A., & Zaidi, A., (2010). A brief review of compressed stabilized earth brick (CSEB). *CSSR 2010: 2010 International Conference on Science and Social Research (Cssr)* (pp. 999–1004). https://doi.org/10.1109/CSSR.2010.5773936 (accessed on 22 February 2020).

36. Safiuddin, M., Salam, M. A., & Jumaat, M. Z., (2011). Utilization of palm oil fuel ash in concrete: A review. *Journal of Civil Engineering and Management, 17,* 234–247. https://doi.org/10.3846/13923730.2011.57445010.1016/S0958-9465(97)00034-6 (accessed on 22 February 2020).

37. Shakir, A. A., Naganathan, S., & Mustapha, K. N., (2013). Properties of bricks made using fly ash, quarry dust and billet scale. *Construction and Building Materials, 41,* 131–138. https://doi.org/10.1016/j.conbuildmat.2012.11.077 (accessed on 22 February 2020).

38. Shyam, P. K., & Rao, C. H., (2016). Study on compressive strength of quarry dust as fine aggregate in concrete. *Advances in Civil Engineering.* https://doi.org/10.1155/2016/1742769 (accessed on 22 February 2020).

39. Simion, H. K., (2009). Design of interlocking bricks for enhanced wall construction flexibility, alignment accuracy, and load bearing. *PhD Thesis.* School of Engineering, The University of Warwick.

40. Sitton, J. D., & Story, B. A., (2016). Estimating soil classification via quantitative and qualitative field testing for use in constructing compressed earth blocks. *Procedia Engineering, 145,* 860–867. https://doi.org/10.1016/j.proeng.2016.04.112 (accessed on 22 February 2020).

41. Sitton, J. D., Zeinali, Y., Heidarian, W. H., & Story, B. A., (2018). Effect of mix design on compressed earth block strength. *Construction and Building Materials, 158,* 124–131. https://doi.org/10.1016/j.conbuildmat.2017.10.005 (accessed on 22 February 2020).

42. Thomas, B. S., Kumar, S., & Arel, H. S., (2017). Sustainable concrete containing palm oil fuel ash as a supplementary cementations material: A review. *Renewable and*

Sustainable Energy Reviews, 80, 550–561. https://doi.org/10.1016/j.rser.2017.05.128 (accessed on 22 February 2020).

43. Thomas, J., & Harilal, B., (2015). Properties of cold bonded quarry dust coarse aggregates and its use in concrete. *Cement and Concrete Composites, 62*, 67–75. https://doi.org/10.1016/j.cemconcomp.2015.05.005 (accessed on 22 February 2020).

44. Tian, G., Wang, W., Mu, B., Kang, Y., & Wang, A., (2015). Facile fabrication of carbon attapulgite composite for bleaching of palm oil. *Journal of the Taiwan Institute of Chemical Engineers, 50*, 252–258. https://doi.org/10.1016/j.jtice.2014.12.021 (accessed on 22 February 2020).

45. Venkatarama, R. B. V., & Prasanna, K. P., (2011). Cement stabilized rammed earth. Part A: Compaction characteristics and physical properties of compacted cement stabilized soils. *Materials and Structures, 44*(3), 681–693. https://doi.org/10.1617/s11527-010-9658-9 (accessed on 22 February 2020).

46. Walker, P. J., (1995). Strength, durability, and shrinkage characteristics of cement stabilized soil blocks. *Cement and Concrete Composites, 17*(4), 301–310. https://doi.org/10.1016/0958-9465(95)00019-9 (accessed on 22 February 2020).

47. Yetgin, S., & Çavdar, A., (2015). Study of effects of natural pozzolan on properties of cement mortars. *Journal of Materials in Civil Engineering, 1561*. https://doi.org/10.1061/(ASCE)0899-1561(2006)18 (accessed on 22 February 2020).

48. Raman, S. N., Ngo, T., Mendis, P., & Mahmud, H. B., (2011). High-strength rice husk ash concrete incorporating quarry dust as a partial substitute for sand. *Construction and Building Materials, 25 (7)*, 3123 – 3130. https://doi.org/10.1016/j.conbuildmat.2010.12.026.

49. Sharom, N., (2016). Performance of Eco Process Pozzolan Foamed Concrete as Cement Replacement. *B.Eng (Hons.) Civil Engineering Thesis, Faculty of Civil Engineering and Earth Resources, Universiti Malaysia Pahang.* http://umpir.ump.edu.my/id/eprint/13018/ (accessed 1 March 2020).

CHAPTER 13

An Overview of Zn(II) Complexes and Device Properties for Organic Light-Emitting Diodes (OLEDs) Applications

JAYDIP SOLANKI and KIRAN R. SURATI

Department of Chemistry, Sardar Patel University,
Vallabh Vidyanagar, Anand, Gujarat–388120, India,
E-mail: kiransurati@yahoo.co.in (K. Surati)

ABSTRACT

The present manuscript review the structures of Zn(II) complexes in the recent past and their use in the field of organic light-emitting diodes (OLEDs) due to their cost-effective, eco-friendly, and suitable chemical nature as emissive and transporting ability. Here we also summarized the device performance properties and their structures to correlate the ability of the molecular structure with device performance.

13.1 INTRODUCTION

Since the first report of the organic light-emitting diode (OLEDs)with tris (8-hydroxyquinoline) aluminum (Alq_3) complex by C. W. Tang at el. in 1987 [1] there are many developments in the materials are device fabrication technology. The huge reports available on luminescence materials with various phenomena such as fluorescence, phosphorescence, and thermally activated delayed fluorescence (TADF). [2–12] Many p-type and n-type of organic semiconductors are reported with different device structures (Figure 13.1). The p-type and n-type materials utilized for OLED device fabrication for lighting and display application (Figures

13.2 and 13.3) [13]. Luminescence materials are also equally important to make the high-performance efficient device, where these materials placed between p and n-type materials and respective cathode and anode. Among all types of luminescence materials Iridium (III) complexes show a significance contribution in the development of display and lighting industries [14]. Iridium based complexes also attract many other applications such as sensor, optoelectrical device such as organic field-effect transistor (OFET), organic photovoltaics (OPV) and many other. These materials having good features like tuning of emission color, higher phosphorescence quantum efficiency (φ_p), tunable PL lifetime (T_p), chemical inertness and good photostability and thermal stabilities, etc. [15–17]. Though these complexes also suffer from the limitation such as high cost of Ir(III) precursors and toxicity. Besides these Ir(III) complexes also observed low synthesis yield [18].

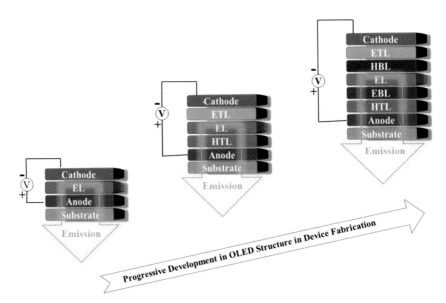

FIGURE 13.1 Progressive development in OLED structure; emissive layer (EL), electron transport layer (ETL), hole transport layer (HTL), electron blocking layer (EBL), hole blocking layer (HBL).

In view of the above fact, it is really an important to design and developed alternate materials which having low cost, environmentally friendly and

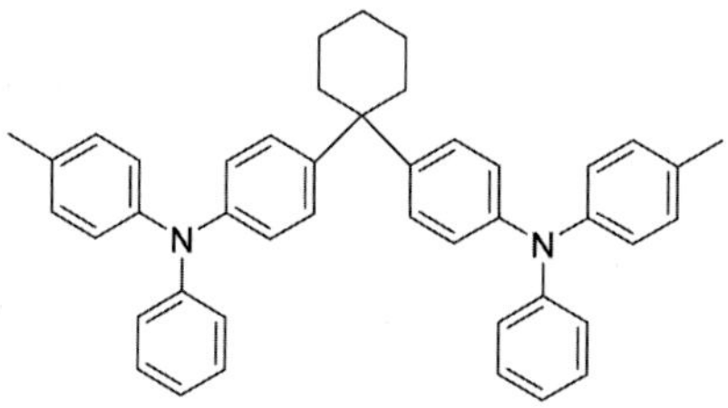

TAPC

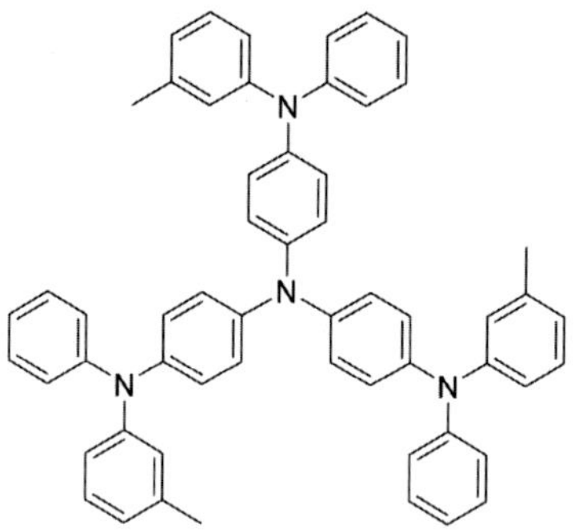

m-MTDATA

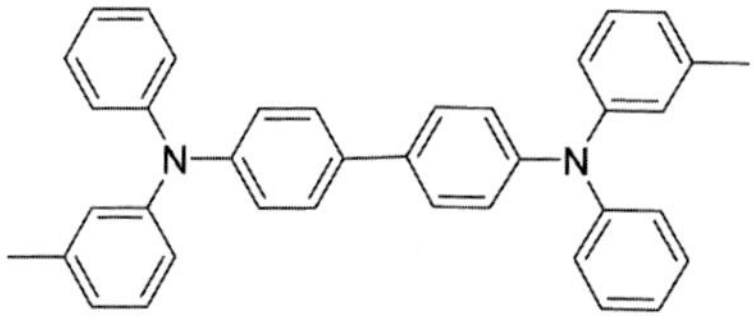

TPD

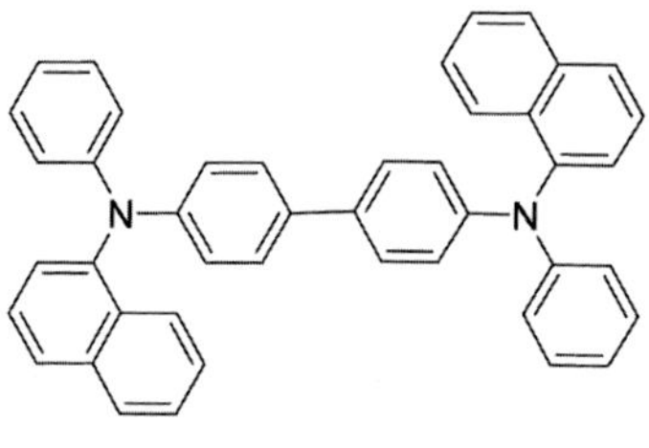

NPB

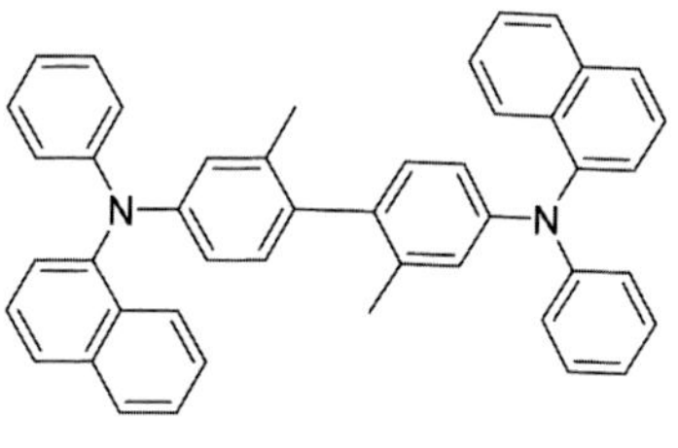

α-NPD

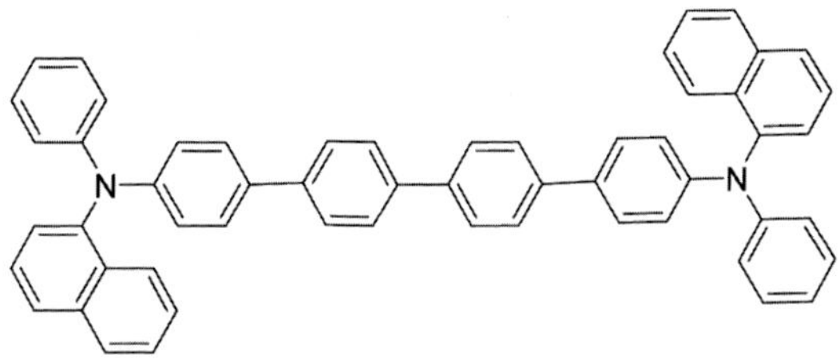

4P-NPD

FIGURE 13.2 p-type materials; 1,1-bis[(di-4-tolylamino)phenyl]cyclohexane (TAPC), N,N'-bis-(1-naphthyl)-N,N'-diphenyl benzidine (α-NPD), tris(4-carbazoyl-9-ylphenyl) amine (TCTA), N,N'-di-1-naphthalenyl-N,N'-diphenyl-[1,1':4,'1":4,"1"'-quaterphenyl]-4,4"'-diamine (4P-NPD), N,N'-bis(I)naphthyl)-N,N'-diphenyl-1,1'-biphenyl-4,4'-diamine (NPB), N,N'-bis(metatolyl)-N,N'-diphenylbenzidine (TPD), 4,4,'4"-tris((3-methylphenyl) phenyl amino)triphenylamine (m-MTDATA).

good synthetic yield. The Zinc(II) complexes are show fluorescence and eco-friendly as compared with heavy metal complexes such as Iridium, Osmium, Platinum, etc. It also significate to note that Zn-precursors is approximately 1400 times cheaper than the standard $IrCl_3 \cdot xH_2O$ precursor and 1000 time cheaper than Os and Pt precursors [18]. This comparison can easily be pushed back one step further with properties of Zn(II) complexes as electron-transporting material in place of conventional purely organic host materials [18, 19]. Due to the above-mentioned reason the present review mainly focuses on characteristic properties of Zn(II) complexes-based device and their structure. It will help to new research for the development of novel materials of Zn(II) complexes with the ability to fine-tune the emission of color, efficiency, and stability. Further, we also provide the account of device fabrication and molecular structures of the last few years to explore the suitability of Zn(II) complexes as the emitter in OLEDs.

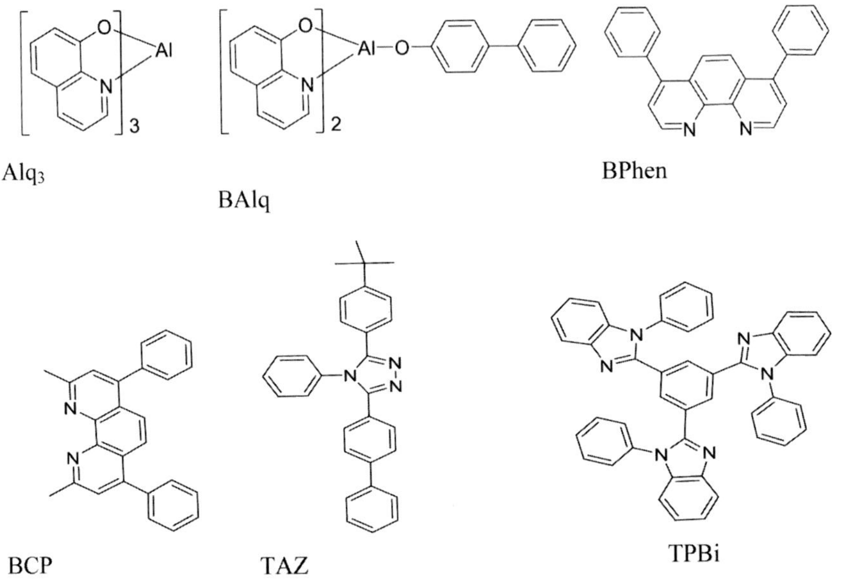

Alq₃   BAlq   BPhen

BCP   TAZ   TPBi

FIGURE 13.3 n-type materials; typical example of ETL materials are tris (8-quinolinolato) aluminum (Alq3), 4-biphenyloxolate aluminum(III) bis(2-methyl-8-quinolinato)-4-phenylphenolate (BAlq), 1,3,5-tri(phenyl-2-benzimidazole)benzene (TPBi), 3-phenyl-4 (10-naphthyl)-5-phenyl-1,2,4-triazole (TAZ), BPhen, and 2,9-dimethyl-4,7-diphenyl-1,10-phenanthroline (BCP).

Since the 2000 large number of publications on Zn(II) complexes for OLEDs application evidence with the diagram (Figure 13.4) continue increasing publications. These publication including 263 articles, 84 conference paper, 02 book chapter and 04 reviews. Here in our discussion on the last few years (after 2014) reports because the review article by F. Dumur already discusses the development of Zn(II) complexes up to 2013 in systematic manner [18]. In the year, 2018 Bizzari and Brase et al. also reported sustainable metal complexes as alternative materials for Ir(III) compound. They mainly review the copper and Zinc complexes and their features like cooperativity, chirality, and printing [20]. The progressive development of Zn(II) complexes were also reflect by their publication details shown in Figure 13.4.

In view of the above, the present review consists structure Zn(II) metal complexes, device structure, and their performance characteristics to understand the properties of the emitters as well as host materials.

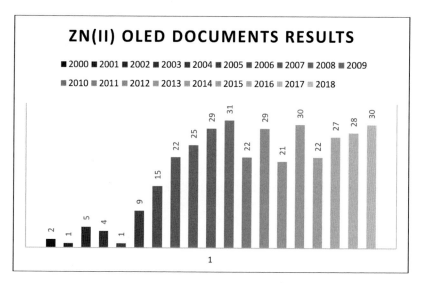

FIGURE 13.4   Zinc based publication year wise.

13.2   OVERVIEW OF ZN(II) COMPLEXES FOR OLEDS

F. Dumur reports the first overview of zinc complexes was used in OLEDs from 2000 to 2014. Particularly, bis(8-hydroxyquinolinato) zinc(II)

Znq$_2$ have strong yellow florescence. Other substitution on quinoline ring was examined to fine-tuning the emission of complexes. In 2013–14, Zn-based heteroleptic complexes of 8-hydroxyquinoline (8-HQ) derivative are (2- (2-hydroxyphenyl) benzoxazolato) (5,7-dimethyl-8-hydroxyquinolinato) zinc(II) Zn(HPB)(Me$_2$q) (C1), (2-(2-hydroxy-phenyl) benzoxa-zolato) (2-carbonitrile-8-hydroxyquinolinato) zinc(II) Zn(HPB)(CNq) (C2) [21], (2-(2-hydroxyphenyl) benzoxazolato) (5-chloro-8-hydroxyquinolinato) zinc(II) Zn(HPB)(Clq) (C3) and (2-(2-hydroxyphenyl) benzoxazolato) (5,7-dichloro-8-hydroxyquinolinato) zinc(II) Zn(HPB)(Clq) (C4)[22]. In the following multilayer device configuration, ITO/ α-NPD/Zinc complexes (C1 to C4)/BCP/Alq$_3$/ LiF/Al, the maximum brightness of 7017, 6925, 6020, 5870 cd/m^2; a current efficiency of 3.6, 2.3, 1.77, 1.92 cd/A; power efficiency of 0.94, 0.32, 0.56, 0.45 lm/W were found for the C1, C2, C3 and C4 respectively. A Schiff base bis (salicylidene)propylene-1,3-diaminato) zinc(II) Zn(salpen)$_2$(C5) was found a maximum brightness 1990 cd/m^2 and current efficiency 0.69 cd/A with above mentioned same device configuration [23]. The green electroluminescence from bis(acetato)-bis (μ3-quinolin-8-olato) tetrakis(μ2-quinolin-8-olato) tetrazinc(II) (C6), Bis(2-methylquinolin-8-olato)-bis[(acetato)-(methanol) zinc(II)] (C7), 8-Hydroxy-2-methylquinolinium dibromido(2-methylquinolin-8-olato) zincate(II) methanol monosolvate(C8) and bromido(quinolin-8-ol) (quinolin-8-olato) zinc(II) methanol monosolvate(C9) were fabricated the multilayer device ITO/PVK: PBD/C6 to C9 (25 nm)/BCP/Alq3/Al, a maximum brightness 1212, 1341, 1025.3, 1113 cd/m^2 for C6 to C9 respectively. In additionally, the zinc complex C6 having green to white electroluminescence with the different thickness in device performance and at 12 nm thickness of C6 have greatest maximum brightness at 4530 cd/m^2, current density of 398.32 mA/cm^2 with the white color coordinates (0.22, 0.36) [24]. Further, the quinaldic acid with 8-HQ ligand-based zinc complexes Zn(HQA) (C10) device performance with ITO/NPB/Zn(HQA)/Liq/Al configuration, a current density 150 mA/ m^2 and maximum brightness 190 cd/m^2 at 15 V [25]. D. Kim et al. reported the two zinc complexes are used as the emitter layer. [bis-(2-(2-hydroxyphenyl) benzoxazole)] zinc(II) Zn(HPB)$_2$ (C11) as blue emitter and (bis 8-hydroxy-7-propylquinoline) zinc(II) Zn(HPQ)$_2$ (C12) as yellow emitter used in device fabrication was ITO/PEDOT: PSS/NPB/ C11/C12/Alq$_3$/LiAl configuration with a maximum luminance 12,000 cd/m^2 and current density 800 mA/cm^2 moreover, the thickness of Zn(HPQ)$_2$(C12) was 20 nm then the device emitted white light [26]. A Schiff base bis{[3-(2-tosylaminobenzylidene) amino] quinolinato} zinc(II) (C13)had poor result compare to bis[2-(*N*-tosylamino) benzylidene-4′-dimethylaminophenylaminato] zinc(II) (C14) [27, 28]. The device performance of C13 was 126 cd/m^2 at 10 V whereas C14 was 1300 cd/m^2 at 14 V a maximum luminescence. After that in 2017, Burlov et al. [29] reported the Schiff base Bis{(4-methyl-N-[2-[(E)-2-pyridy-liminomethyl] phenyl)] benzene sulfonamide} zinc(II) (C15) with same device configuration ITO/CuPc/2-TNATA/TPD/C15/BCP/Bphen/LiF/Al give more than 1000 cd/m^2 luminance [29]. In comparison of Schiff, base C14 and C15 based devices proved to exhibit greater performance than C13. The tetranuclear zinc complexes of substituted 7-azaindoles, [Zn$_4$O(AID-4-Phenyl)$_6$] (C16), [Zn$_4$O(AID-4-Thiophen-2-yl)$_6$] (C17) and [Zn$_4$O(AID-4-Naphthalen-2-yl)$_6$] (C18) are the performance of OLED device with ITO/PEDOT: PSS/PVK: OXD-7:C16-C18/TmPyPb/TPBi/LiF/Al configuration and C16 to C18 complexes used the dopant concentration of 2–20 wt% as blue dopant. When dopant concentration was 8wt% of C16 than device showed the highest EQE 5.55% compare to C17 and C18 [30]. Jafari et al. introduce π-electron-rich host material bis(acetato)-bis(µ3-quinolin-8-olato) tetrakis(µ2-quinolin-8-olato) tetrazinc(II) (C6) with 2wt% of meso-tTetraphenylporphyrin TPP (C19), meso-tetrakis (4-dimethoxyphenyl) porphyrin as 4-TPP (C20), meso-tetrakis (3,4-dimethoxyphenyl) porphyrin as 3-4TPP (C21), and tetraphenylporphyrinzinc as ZnTPP(C22) fabricated the device ITO/PEDOT: PSS/PVK/C6: 2wt% C19 to C22/Al, among the four C19-C22, C22 proved to be the best result with 3500 cd/m^2 and C6 was promising host material for red OLEDs [31]. M. Janghouri introduced the concept that the modification of size of emissive dopant could change the optical and electrical properties, on that basis he synthesized the four Nano zinc(II) complexes with different sizes 48, 60, 76 and 87 nm are referred with labels C23, C24, C25 and C26, the fabricated the following device ITO/PEDOT: PSS (45 nm)/PVK (30 nm)/C23-C26 (90 nm) PBD (30 nm)/Al (150 nm). Among the four different sizes of Nano zinc complexes, 60 nm (C24) showed the highest a maximum luminance of 7492 cd/m^2 and maximum external quantum efficiencies (EQE) of 2.1% [32]. Zn(DFP-SAMQ)$_2$(C27), Zn(PSA-BTZ)$_2$(C28), Zn(POPS-BTZ)$_2$(C29) showed the maximum brightness 140, 230 and 270 cd/m^2 respectively [33].

2-(1-phenyl-1H-phenanthro[9,10-d] imidazol-2-yl) phenol)zinc(II) $Zn(PPI)_2$ (C30) combined with dopant $Ir(piq)_3$ and $Ir(pmiq)_2(acac)$ deep red phosphorescent organic light-emitting diodes (PHOLEDs) are fabricated the following device $ITO/MoO_3/NPB/TCTA/Zn(PPI)_2$: 8 wt% $Ir(piq)_3$ or $Ir(pmiq)_2(acac)/TPBi/LiF/Al$ showed a maximum luminance more than 10,000 cd/m^2 and EQE of 9.5% for Ir(piq)3 based device and 13.1% for Ir(pmiq)2(acac) device [34]. Zn(II) complexes based on bis-Schiff base ligands, bis-Zn(II) salphen complexes with pyridyl functionalized ligands were Zn-N (C31), Zn-S (C32), Zn-P (C33) and Zn-Si (C34). From the photoluminescence, the study investigates the films for these C31-C34 complexes indicated that the degree of molecular aggregation to enhance their emission intensity has reduced and advantage of the charge carrier injection/transporting ability of the pyridyl functionalized ligands [35]. The OLED device fabricated with the following configuration of ITO/PEDOT: PSS/C31-C34 (6–10 wt%): CBP/TPBi/ LiF/Al, comparing the four complexes C31–34 device with different doping level, the C31 showed to be best emitter and the significant luminescence (L_{max}) of 3589 cd/m^2, EQE of 1.46%, current efficiency of 4.1 cd/A and power efficiency of 3.8 lm/W [35]. The first example of novel class of aggregation-induced emission (AIE) materials as $Zn_2(AMOX)_4$ where (AMOX =4-bromo-*N, N'*-diphenylbenzamidinate*N*-oxide) (C35) was fabricated the white-green OLED with following device configuration, ITO/PEDOT: PSS/PVK-h/PVK-l: PBD: C35(3–12 wt.%)/BCP/ Alq_3/Al/Ag, the device contains 3 wt% of C35 obtained the best result the 1.12 cd/A current density (LE) and 0.30 lm/W power efficiency (PE) compare to other device and further the value of LE and PE are low in comparison to those reported for phosphoresce and TADF-based WOLEDs [36] (Table 13.1).

TABLE 13.1 Device Structure of Zinc Complexes with Different Properties

Complex	Device Structure	V_{ON} (V)	Max η_c (cd/A)	Max η_p (lm/W)	Max Brightness (cd/m²)	CIE Coordinate	EQE (%)	λ_{EL} (nm)	References
C1	ITO/α-NPD (30 nm)/Zn(HPB)(Me$_2$q) (35 nm)/BCP(6 nm)/Alq3 (28 nm)/LiF (1 nm)/Al (100 nm)	10.0	3.6	0.94	7017	(0.48, 0.50)		567	[21]
C2	ITO/α-NPD (30 nm)/Zn(HPB)(CNq) (35 nm)/BCP(6 nm)/Alq3 (28 nm)/LiF (1 nm)/Al (100 nm)	7.0	2.3	0.32	6925	(0.48, 0.49)		569	[21]
C3	ITO/α-NPD (30 nm)/Zn(HPB)(Clq) (35 nm)/BCP(6 nm)/Alq3 (28 nm)/LiF (1 nm)/Al (100 nm)	7.0	1.77	0.56	6020	(0.37, 0.57)		549	[22]
C4	ITO/α-NPD (30 nm)/Zn(HPB)(Cl$_2$q) (35 nm)/BCP(6 nm)/Alq3 (28 nm)/LiF (1 nm)/Al (100 nm)	6.0	1.92	0.45	5870	(0.40, 0.56)		554	[22]
C5	ITO/α-NPD (30 nm)/Zn(Salpen)$_2$ (35 nm)/BCP(6 nm)/ Alq3 (28 nm)/LiF (1 nm)/Al (100 nm)	7.0	0.69	0.26	1990	(0.26, 0.50)		513	[23]
C6	ITO/PVK:PBD (50 nm)/C6 (25 nm)/BCP(5 nm)/ Alq3(25 nm)/Al(180 nm)	9.0			1212	(0.36, 0.54)		535	[24]
C7	ITO/PVK:PBD (50 nm)/C7 (25 nm)/BCP(5 nm)/ Alq3(25 nm)/Al(180 nm)	9.0	0.77		1341.41	(0.29,0.51)		519	[24]
	ITO/PVK:PBD (50 nm)/C7 (20 nm)/BCP(5 nm)/ Alq3(25 nm)/Al(180 nm)	9.0	0.86		1673.12	(0.25, 0.46)			[24]
	ITO/PVK:PBD (50 nm)/C7(18 nm)/BCP(5 nm)/ Alq3(25 nm)/Al(180 nm)	9.0	0.87		2227.925	(0.25, 0.40)			[24]
	ITO/PVK:PBD (50 nm)/C7 (12 nm)/BCP(5 nm)/ Alq3(25 nm)/Al(180 nm)	9.0	1.3		4530	(0.36, 0.54)			[24]
C8	ITO/PVK:PBD (50 nm)/C8 (25 nm)/BCP(5 nm)/ Alq3(25 nm)/Al(180 nm)	9.0			1025.3	(0.19, 0.45)		493	[24]

TABLE 13.1 *(Continued)*

Complex	Device Structure	V_{ON} (V)	Max η_c (cd/A)	Max η_p (lm/W)	Max Brightness (cd/m²)	CIE Coordinate	EQE (%)	λ_{EL} (nm)	References
C9	ITO/PVK:PBD (50 nm)/C9 (25 nm)/BCP(5 nm)/Alq3(25 nm)/Al(180 nm)	9.0			1113	(0.38, 0.50)		548	[24]
C10	ITO/NPB(50 nm)/Zn(HQA) (50 nm)/Liq (5 nm)/Al (120 nm)	10.0	0.15		190			550	[25]
C11 and C12	ITO/PEDOT:PSS (23 nm)/NPB (40 nm)/Zn(HPB)2 (40 nm)/Zn(HPQ)2(20 nm)/Alq3(10 nm)/LiAl (120 nm).	4.75	0.8		12,000	(0.319, 0.338)		450,530	[26]
C13	ITO/CuPc(5 nm)/2-TNATA(35 nm)/TPD(10 nm)/C13(20 nm)/BCP(5 nm)/Bphen(20 nm)/LiF(1 nm)/Al (100 nm)	10.0			126	(0.436, 0.492)		569	[27]
C14	ITO/PEDOT:PSS(40 nm)/2-TNATA(27 nm)/TPD(6 nm)/C14(20 nm)/BCP(5 nm)/Bphen(22 nm)/LiF(0.9 nm)/Al (100 nm)				1300	(0.431, 0.537)		555	[28]
C15	ITO/CuPc(5 nm)/2-TNATA(35 nm)/TPD(6.5 nm)/C15(23 nm)/BCP(5 nm)/Bphen(26 nm)/LiF(0.8 nm)/Al (100 nm)	4.0	0.6	1.0	1000	(0.409, 0.506)		525,650	[29]
C16	ITO/PEDOT:PSS/PVK:OXD-7:C16 (2%)/TmPyPb/TPBi/LiF/Al		4.29	2.31	1080	(0.16, 0.15)	3.94		[30]
	ITO/PEDOT:PSS/PVK:OXD-7:C16 (8%)/TmPyPb/TPBi/LiF/Al		8.12	4.28	6503	(0.16, 0.19)	5.55		[30]
	ITO/PEDOT:PSS/PVK:OXD-7:C16 (20%)/TmPyPb/TPBi/LiF/Al		5.12	2.45	3480	(0.16, 0.21)	3.18		[30]

TABLE 13.1 *(Continued)*

Complex	Device Structure	V_{ON} (V)	Max η_c (cd/A)	Max η_p (lm/W)	Max Brightness (cd/m²)	CIE Coordinate	EQE (%)	λ_{EL} (nm)	References
C17	ITO/PEDOT:PSS/PVK:OXD-7:C17 (2%)/TmPyPb/TPBi/LiF/Al		4.14	2.37	2290	(0.17, 0.20)	2.83		[30]
	ITO/PEDOT:PSS/PVK:OXD-7:C17 (8%)/TmPyPb/TPBi/LiF/Al		6.00	3.14	6300	(0.18, 0.26)	3.33		[30]
	ITO/PEDOT:PSS/PVK:OXD-7:C17 (2%)/TmPyPb/TPBi/LiF/Al		3.00	1.50	4095	(0.19, 0.30)	1.49		[30]
C18	ITO/PEDOT:PSS/PVK:OXD-7:C18 (2%)/TmPyPb/TPBi/LiF/Al		2.14	1.12	1240	(0.17, 0.16)	1.92		[30]
	ITO/PEDOT:PSS/PVK:OXD-7:C18 (8%)/TmPyPb/TPBi/LiF/Al		5.57	3.00	4650	(0.17, 0.21)	3.61		[30]
	ITO/PEDOT:PSS/PVK:OXD-7:C18 (20%)/TmPyPb/TPBi/LiF/Al		4.12	1.88	5430	(0.18, 0.25)	2.29		[30]
C19	ITO/PEDOT: PSS (50 nm)/PVK (60 nm)/C6: C19 (45 nm)/Al(150 nm)	6.8	1.1		1042	(0.39, 0.45)			[31]
C20	ITO/PEDOT: PSS (50 nm)/PVK (60 nm)/C6: C20 (45 nm)/Al(150 nm)	5.9	1.7		1770	(0.60, 0.36)			[31]
C21	ITO/PEDOT: PSS (50 nm)/PVK (60 nm)/C6: C21 (45 nm)/Al(150 nm)	8.7	0.6		651	(0.70, 0.28)			[31]
C22	ITO/PEDOT: PSS (50 nm)/PVK (60 nm)/C6: C22 (45 nm)/Al(150 nm)	5.0	1.78		3500	(0.34, 0.54)			[31]
C23	ITO/PEDOT: PSS (45 nm)/PVK (30 nm)/C23 (90 nm) PBD (30 nm)/Al(150 nm)	9.3	1.5	0.94	2598	(0.28, 0.55)	0.8		[32]

TABLE 13.1 *(Continued)*

Complex	Device Structure	V_{ON} (V)	Max η_c (cd/A)	Max η_p (lm/W)	Max Brightness (cd/m²)	CIE Coordinate	EQE (%)	λ_{EL} (nm)	References
C24	ITO/PEDOT: PSS (45 nm)/PVK (30 nm)/C24 (90 nm) PBD (30 nm)/Al(150 nm)	6.7	2.48	1.16	7492	(0.33, 0.41)	2.1	541	[32]
C25	ITO/PEDOT: PSS (45 nm)/PVK (30 nm)/C25 (90 nm) PBD (30 nm)/Al(150 nm)	8.4	1.6	0.96	3086	(0.31, 0.54)	1.1		[32]
C26	ITO/PEDOT: PSS (45 nm)/PVK (30 nm)/C26 (90 nm) PBD (30 nm)/Al(150 nm)	7.2	1.86	1.09	4485	(0.32, 0.59)	1.5		[32]
C27	ITO/PEDOT:PSS/NPD/Zn(DFP-SAMQ)₂/Ca/Al		0.035		140			518	[33]
C28	ITO/PEDOT:PSS/NPD/CBP/Zn(PSA-BTZ)₂/LiF/Al		0.040		230			500	[33]
C29	ITO/PEDOT:PSS/NPD/Zn(POPS-BTZ)₂/LiF/Al		0.070		270			525	[33]
C30	ITO/MoO3 (10 nm)/NPB (50 nm)/TCTA (5 nm)/Zn(PPI)₂: 8 wt%.Ir(piq)₃ (30 nm)/TPBi (40 nm)/LiF (1 nm)/Al	2.7	5.35	6.21	10,000	(0.68, 0.32)	9.5	636	[34]
C31	ITO/PEDOT:PSS (45 nm)/Zn-N (6 wt%):CBP (40 nm)/TPBi (45 nm)/LiF (1 nm)/Al (100 nm)	3.4	4.1	3.8	3589	(0.52, 0.47)	1.46	576	[35]
	ITO/PEDOT:PSS (45 nm)/Zn-N (8 wt%):CBP (40 nm)/TPBi (45 nm)/LiF (1 nm)/Al (100 nm)	3.8	3.0	2.5	3658		1.07		[35]
	ITO/PEDOT:PSS (45 nm)/Zn-N (10 wt%):CBP (40 nm)/TPBi (45 nm)/LiF (1 nm)/Al (100 nm)	3.8	2.5	2.1	2466		0.90		[35]
C32	ITO/PEDOT:PSS (45 nm)/Zn-S (6 wt%):CBP (40 nm)/TPBi (45 nm)/LiF (1 nm)/Al (100 nm)	3.6	1.4	1.2	1627	(0.51, 0.47)	0.49	572	[35]
	ITO/PEDOT:PSS (45 nm)/Zn-S (8 wt%):CBP (40 nm)/TPBi (45 nm)/LiF (1 nm)/Al (100 nm)	3.5	3.3	2.9	2021		1.19		[35]

TABLE 13.1 (*Continued*)

Complex	Device Structure	V_{ON} (V)	Max η_c (cd/A)	Max η_p (lm/W)	Max Brightness (cd/m²)	CIE Coordinate	EQE (%)	λ_{EL} (nm)	References
C33	ITO/PEDOT:PSS (45 nm)/Zn-S (10 wt%):CBP (40 nm)/TPBi (45 nm)/LiF (1 nm)/Al (100 nm)	3.6	2.4	2.1	1734		0.87		[35]
	ITO/PEDOT:PSS (45 nm)/Zn-P (6 wt%):CBP (40 nm)/TPBi (45 nm)/LiF (1 nm)/Al (100 nm)	3.6	3.1	2.7	2086	(0.55, 0.45)	1.18	576	[35]
	ITO/PEDOT:PSS (45 nm)/Zn-P (8 wt%):CBP (40 nm)/TPBi (45 nm)/LiF (1 nm)/Al (100 nm)	3.7	2.4	2.0	1450		0.90		[35]
	ITO/PEDOT:PSS (45 nm)/Zn-P (10 wt%):CBP (40 nm)/TPBi (45 nm)/LiF (1 nm)/Al (100 nm)	3.9	1.8	0.9	1047		0.68		[35]
C34	ITO/PEDOT:PSS (45 nm)/Zn-Si (6 wt%):CBP (40 nm)/TPBi (45 nm)/LiF (1 nm)/Al (100 nm)	3.5	1.7	1.5	1825	(0.53, 0.46)	0.65	576	[35]
	ITO/PEDOT:PSS (45 nm)/Zn-Si (8 wt%):CBP (40 nm)/TPBi (45 nm)/LiF (1 nm)/Al (100 nm)	3.6	2.7	2.3	1985		1.00		[35]
	ITO/PEDOT:PSS (45 nm)/Zn-Si (10 wt%):CBP (40 nm)/TPBi (45 nm)/LiF (1 nm)/Al (100 nm)	3.6	2.1	1.8	2225		0.78		[35]
C35	ITO/PEDOT:PSS (30 nm)/PVK-h (15 nm)/PVK-l:PBD: C35(3 wt.%) (30 nm)/BCP(10 nm)/Alq3(40 nm)/Al(150 nm)/Ag (50 nm)	10.0	1.12	0.30		(0.28, 0.51)		400–450	[36]
	ITO/PEDOT:PSS (30 nm)/PVK-h (15 nm)/PVK-l:PBD: C35(6 wt.%) (30 nm)/BCP(10 nm)/Alq3(40 nm)/Al(150 nm)/Ag (50 nm)	10.0	1.03	0.25		(0.26, 0.53)			[36]
	ITO/PEDOT:PSS (30 nm)/PVK-h (15 nm)/PVK-l:PBD: C35(12 wt.%) (30 nm)/BCP(10 nm)/Alq3(40 nm)/Al(150 nm)/Ag (50 nm)	10.0	0.49	0.13		(0.27, 0.54)			[36]

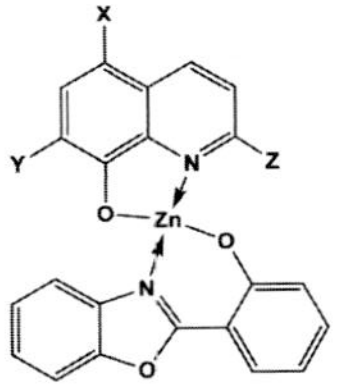

C5: Zn(Salpen)₂

C8

C11: Zn(HPQ)₂

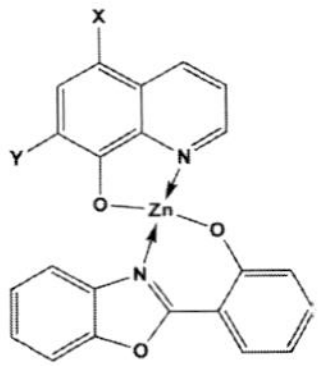

C3: Zn(HPB)(Clq); X=Cl, Y=H
C4: Zn(HPB)(Cl₂q); X=Cl, Y=Cl

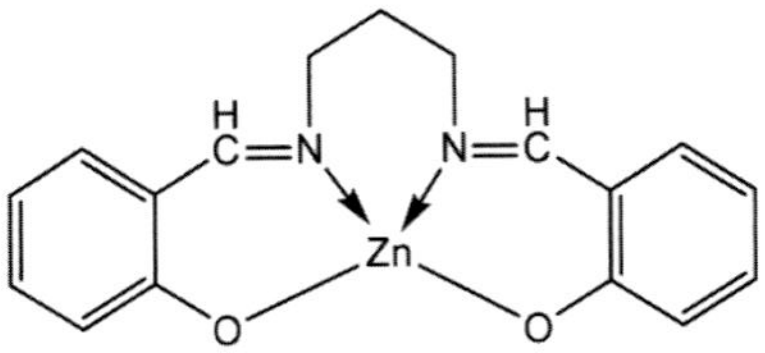

C7

C10: Zn(HQA)

C1: Zn(HPB)(Me₂q); X=Me, Y=Me, Z=H
C2: Zn(HPB)(CNq); X=H, Y=H, Z=CN

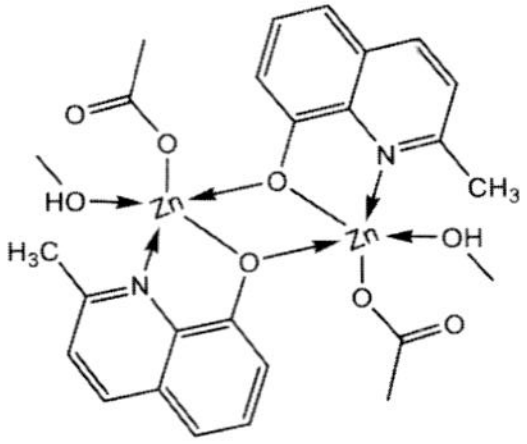

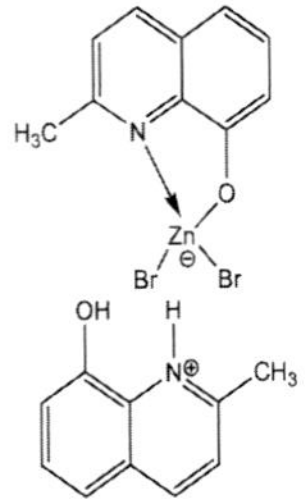

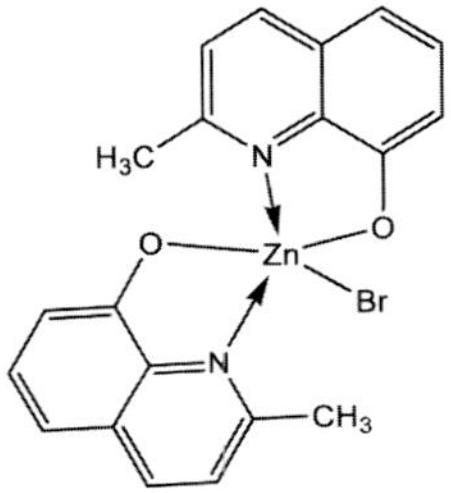

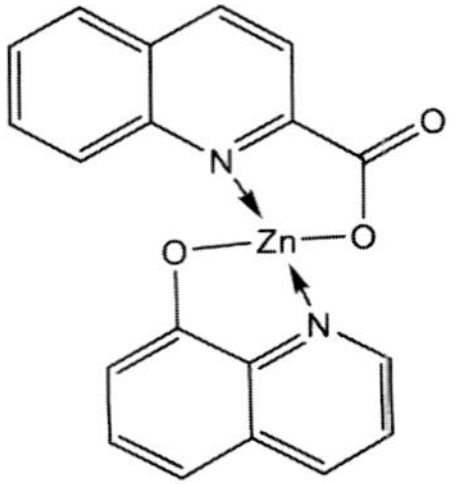

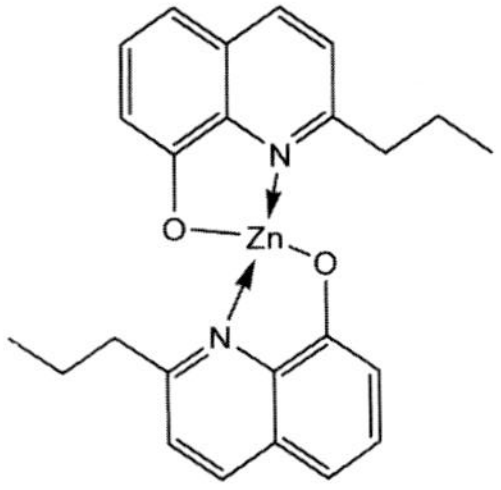

C6

C9

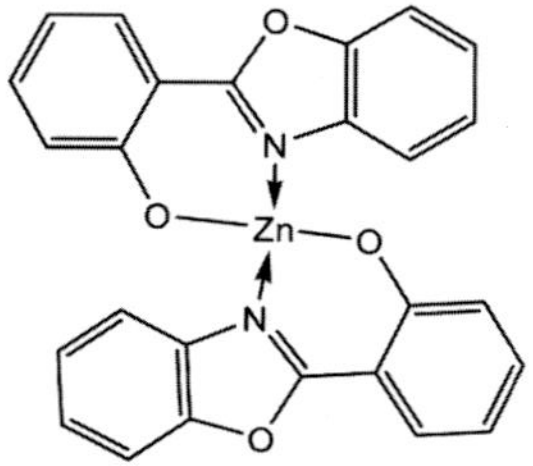 **C12**: Zn(HPB)$_2$	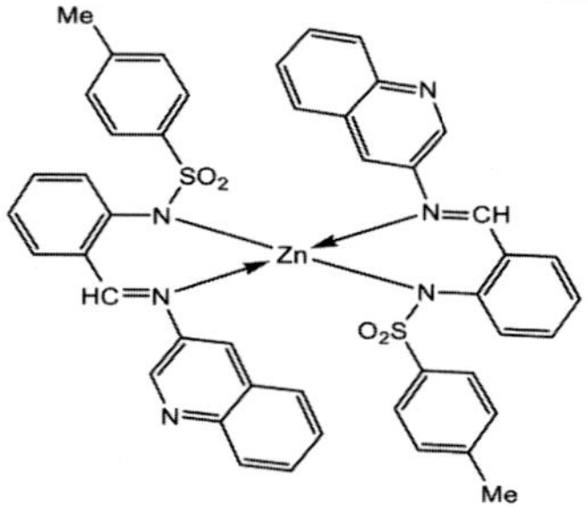 **C13**	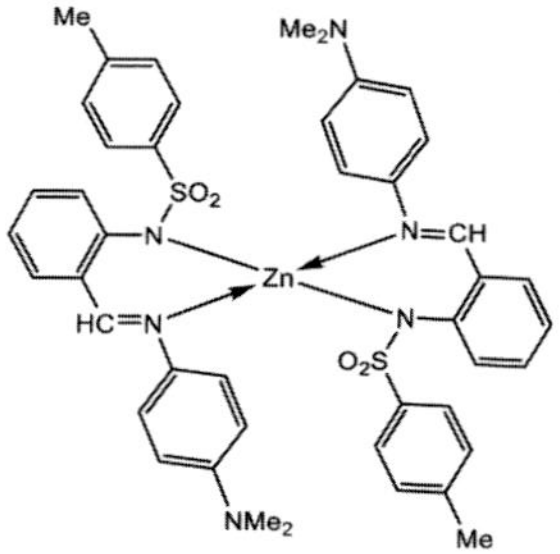 **C14**
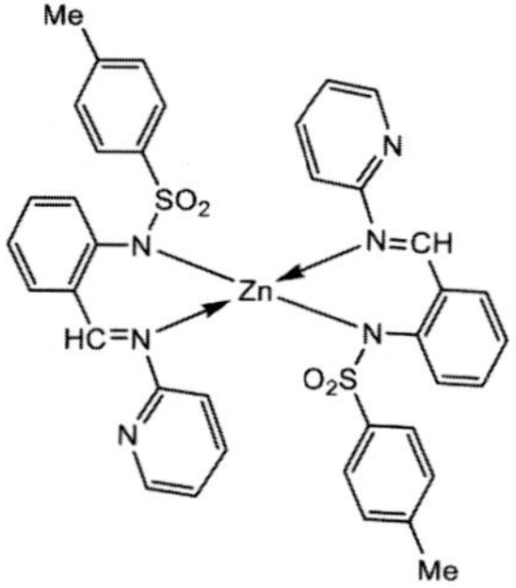 .2CHCl$_3$ **C15**	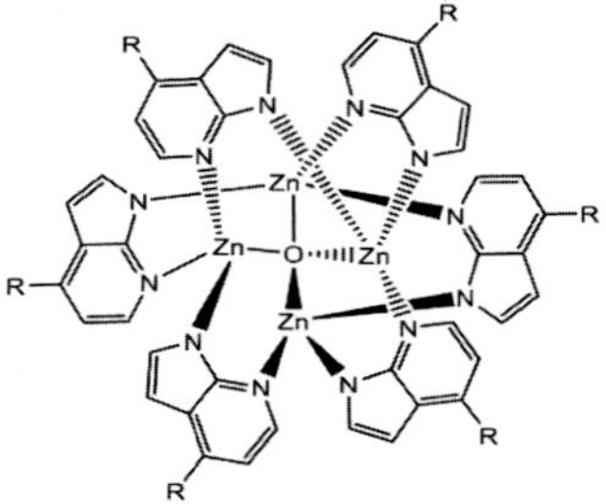	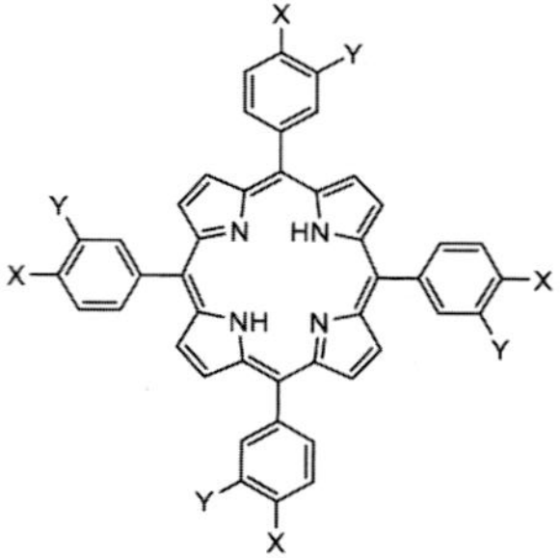 **C19:** X=H, Y=H; **(TPP)** **C20:** X=H, Y=OCH$_3$; **(4-TPP)** **C21:** X= OCH$_3$, Y=OCH$_3$; **(3-4TPP**

	C16	C17	C18
R	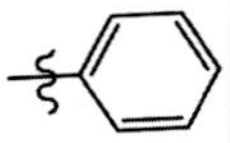	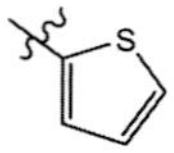	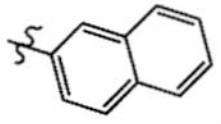

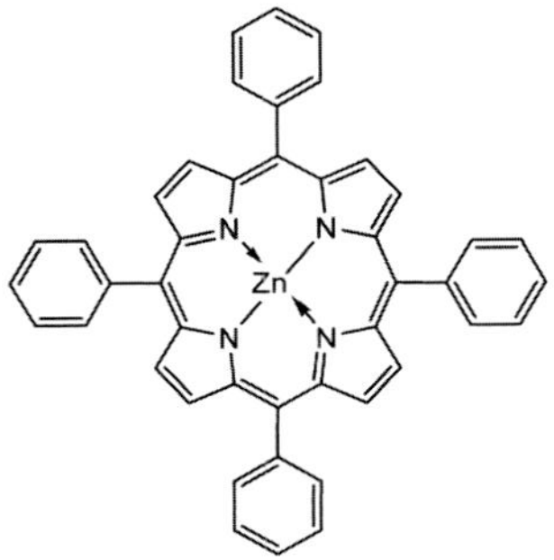

C27: Zn(DFP-SAMQ)$_2$

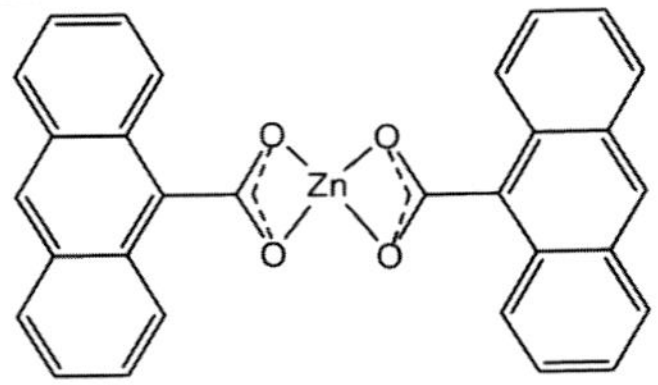

C23: Nano Zncomplex size (48 nm)
C24: Nano Zncomplex size (60 nm)
C25: Nano Zncomplex size (76 nm)
C26: Nano Zncomplex size (87 nm)

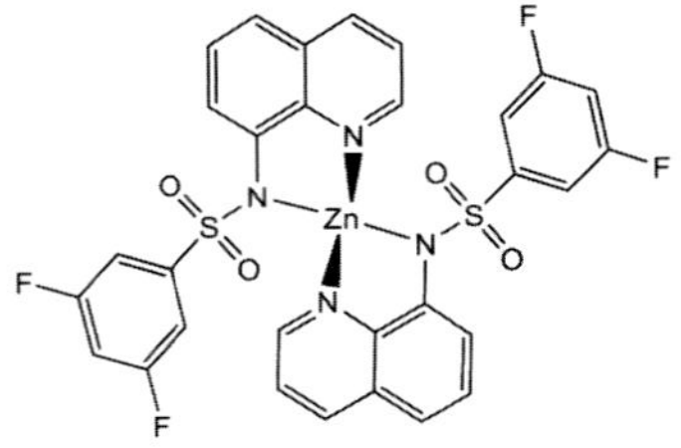

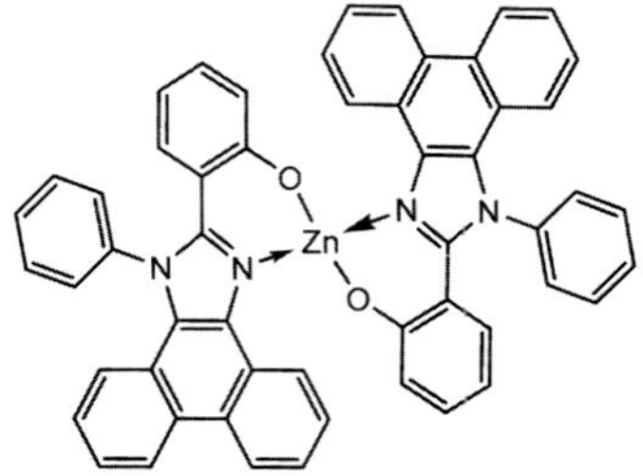

C30: Zn(PPI)$_2$

C22: Zn(TPP)

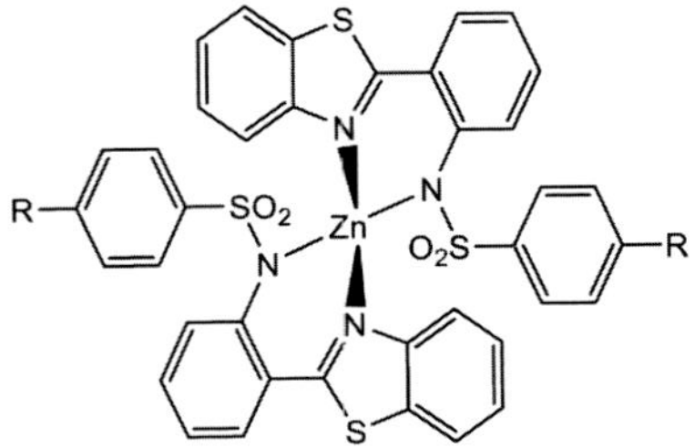

C28: R=H, Zn(PSA-BTZ)$_2$
C29: R=OC$_{15}$H$_{31}$, Zn(POPS-BTZ)$_2$

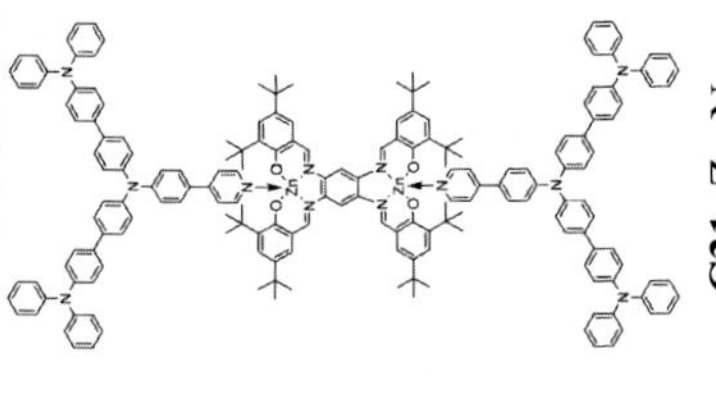

C33: Zn-P

C32: Zn-S

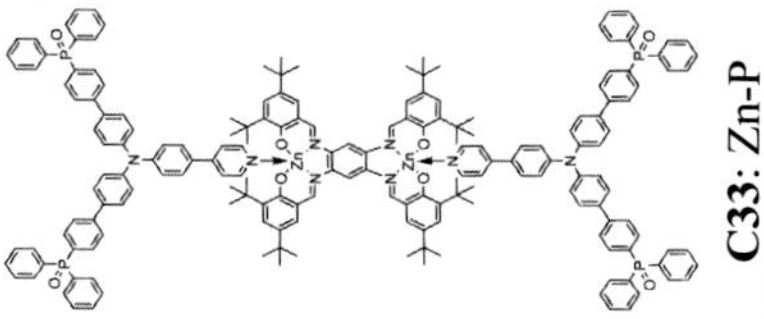

C35

C31: Zn-N

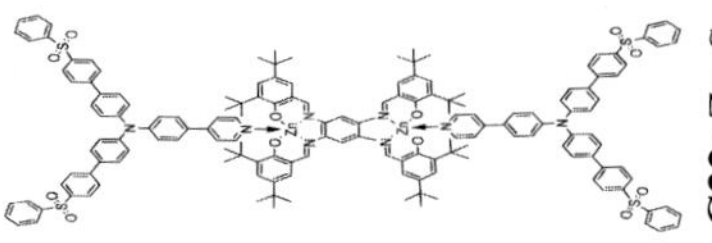

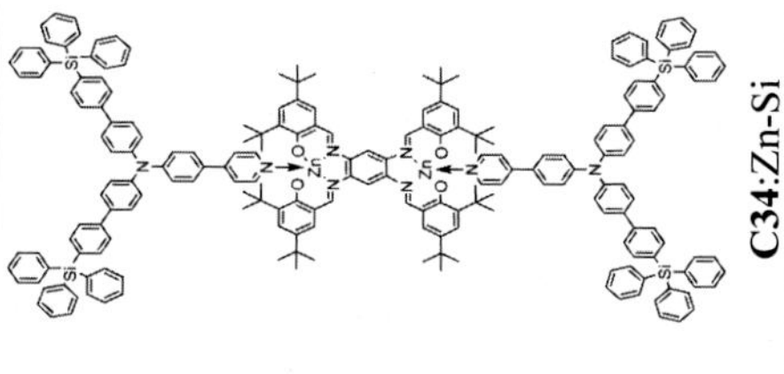

C34:Zn-Si

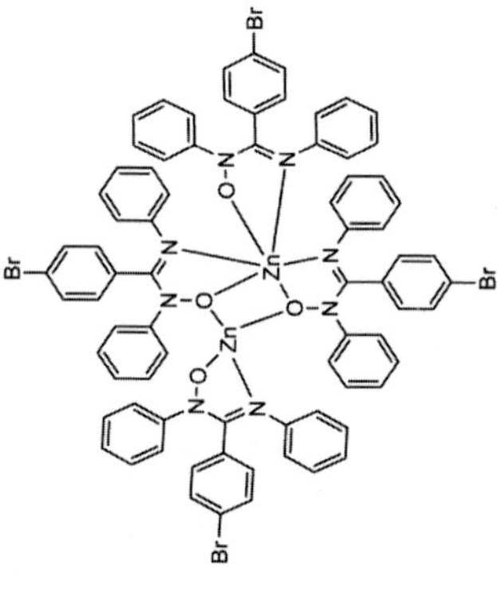

ACKNOWLEDGMENTS

Authors express sincere thanks to the Department of Science and Technology (DST), New Delhi, for financial support under Scheme of Young Scientist and Technologist (SYST) (Ref. No.SP/YO/008/2014(G)) Author (JS) also thankful for project fellowship under this project.

KEYWORDS

- **molecular structure**
- **organic field-effect transistor**
- **organic light-emitting diodes (OLEDs)**
- **performance characteristics**
- **thermally activated delayed fluorescence**
- **Zn(II) complexes**

REFERENCES

1. Tang, C. W.; Van Slyke, S. A., (1987). Organic Electroluminescent Diodes. *Appl. Phys. Lett., 51,* 913–915.
2. Fan, L. J.; Zhang, Y.; Murphy, C. B.; Angell, S. E.; Parker, M. F. L.; Flynn, B. R.; Jones, W. E., (2009). Fluorescent Conjugated Polymer Molecular Wire Chemosensors for Transition Metal Ion Recognition and Signaling. *Coord. Chem. Rev., 253,* 410–422.
3. Tagarea, J.; Vaidyanathan, S., (2018). Recent development of phenanthroimidazole-based fluorophores for blue organic light-emitting diodes (OLEDs): an overview. *J. Mater. Chem. C, 6,* 10138–10173.
4. Godumala, M.; Choi, S.; Choa, M. J.; Choi, D. H., (2019). Recent breakthroughs in thermally activated delayed fluorescence organic light emitting diodes containing non-doped emitting layers. *J. Mater. Chem. C, 7,* 2172–2198.
5. Pereira, L.(2012). *Organic Light Emitting Diodes: The Use of Rare Earth and Transition Metals,* CRC Press, ISBN 9789814267298.
6. Fu, H. S.; Cheng, Y. M.; Chou, P. T.; Chi, Y., (2011). Feeling blue? Blue phosphors for OLEDs. *Mater. Today, 14,* 472–479.
7. Yang, Z.; Mao, Z.; Xie, Z.; Zhang, Y.; Liu, S.; Zhao, J.; Xu, J.; Chi, Z.; Aldred, M. P., (2017). Recent Advances in Organic Thermally Activated Delayed Fluorescence Materials. *Chem. Soc. Rev., 46,* 915−1016.
8. Sree, V. G.; Cho, W.; Shin, S.; Lee, T.; Gal, Y.-S.; Song, M.; Jin, S.-H., (2017). Highly Efficient Solution-Processed Deep-Red Emitting Heteroleptic

Thiophene-Phenylquinoline Based Ir(III) Complexes for Phosphorescent Organic Light-Emitting Diodes. *Dyes Pigm., 139,* 779–787.

9. Lee, Y. H.; Park, S.; Oh, J.; Shin, J. W.; Jung, J.; Yoo, S.; Lee, M. H., (2017). Rigidity-Induced Delayed Fluorescence by Ortho Donor Appended Triarylboron Compounds: Record-High Efficiency in Pure Blue Fluorescent Organic Light-Emitting Diodes. *ACS Appl. Mater. Interfaces, 9,* 24035–24042

10. Wu, R.; Liu, W.; Zhou, L.; Li, X.; Chena, K.; Zhang, H., (2019). Highly efficient green single-emitting layer phosphorescent organic light-emitting diodes with an iridium(iii) complex as a hole-type sensitizer. *J. Mater. Chem. C, 7,* 2744–2750.

11. Sahin, O.; Cinar, M. E.; Tekin, E.; Mucur, S. P.; Topal, S.; Suna, G.; Eroglu, M. S.; Ozturk, O., (2017). White Light Emitting Polymers Possessing Thienothiophene and Boron Units. *Chemistry Select, 2,* 2889–2894.

12. Kumar, S.; Surati, K. R.; Lawrence, R.; Vamja, A. C.; Yakunin, S.; Kovalenko, M. V.; Santos, E. J.G.; Shih, C., (2017). Design and Synthesis of Heteroleptic Iridium(III) Phosphors for Efficient Organic Light-Emitting Devices. *Inorg. Chem., 56,* 15304–15313.

13. Vamja, A. C., (2017). Synthesis and Characterization of Phosphorescent Mixed Ligand Complexes for Organic Light Emitting Diodes, *Doctoral Thesis.*

14. OLED info, April 2019. (https://www.oled-info.com/)

15. Yersin, H.; Rausch, A. F.; Czerwieniec, R.; Hofbeck, T.; Fischer, T., (2011). The triplet state of organo-transition metal compounds. Triplet harvesting and singlet harvesting for efficient OLEDs. *Coord. Chem. Rev., 255,* 2622–2652.

16. Wong, W. Y.(2015). *Organometallics and related molecules for energy conversion,* Springer, ISBN: 978-3-662-46053-5.

17. Schmidt, T. D.; Lampe, T.; Sylvinson M. R., D.; Djurovich, P. I.; Thompson, M. E.; Brütting, W., (2017). Emitter Orientation as a Key Parameter in Organic Light-Emitting Diodes. *Phys. Rev. Applied, 8,* 037001.

18. Dumur, F., (2014). Zinc complexes in OLEDs: An overview. *Synth. Metals, 195,* 241–251.

19. Dumur, F.; Beouch, L.; Tehfe, M.; Contal, E.; Lepeltier, M.; Wantz, G.; Graff, B.;Goubard, F.; Mayer, C. R.; Lalevée, J.; Gigmes, D., (2014). Low-cost zinc complexes for white organic light-emitting devices. *Thin Solid Films, 564,* 351–360.

20. Bizzarri, C.; Spuling, E.; Knoll, D. M.; Volz, D.; Bräseac, S., (2018). Sustainable metal complexes for organic light-emitting diodes (OLEDs). *Coord. Chem. Rev., 373,* 49–82.

21. Kumar, A.; Palai, A. K.; Shrivastava, R.; Kadyan, P. S.; Kamalsanan, M. N.; Singh, I., (2014). n-Type ternary zinc complexes: Synthesis, physicochemical properties and organic light emitting diodes application. *J. Organomet. Chem., 756,* 38–46.

22. Kumar, A.; Srivastava, R.; Kumar, A.; Nishal, V.; Kadyan, P. S.; Kamalasanan, M. N.; Singh, I., (2013). Ternary zinc complexes as electron transport and electroluminescent materials. *J. Organomet. Chem.,740,* 116–122.

23. Nishal, V.; Singh, D.; Kumar, A.; Tanwar, V.; Singh, I.; Srivastava, R.; Kadyan, P. S., (2014). A new zinc–Schiff base complex as an electroluminescent material. *J. Org. Semicond., 2*(1), 15–20.

24. Janghouri, M.; Mohajerani, E.; Amini, M. M.; Najafi, E., (2014). Green–white electroluminescence and green photoluminescence of zinc complexes. *J. Lumin., 154,* 465–474.

25. Im,Y.-H.; Kang, E.; Kim, I. -H.; Kim,D.-E.; Shin, H.-K.; Lee, B.-J., (2014). Synthesis and OLED Properties of Zinc Complexes Based on Quinaldic Acid. *Mol. Cryst. Liq. Cryst. 599:* 1, 105–111.

26. Kim, D. E.; Kwon, Y. S.; Shin, H. K., (2015). Fabrication of White Organic Light Emitting Diode Using Two Types of Zn-Complexes as an Emitting Layer. *J Nanosci. Nanotechnol., 15*(1), 488–491.

27. Burlov, A. S.; Chesnokov, V. V.; Vlasenko, V. G.; Garnovskii, D. A.; Mal´tsev, E. I.; Dmitriev, A. V.; Lypenko, D. A.; Borodkin, G. S.; Revinskiic, Y. V., (2014). Synthesis, structure, and spectral studies of zinc and cadmium complexes with 2-tosylaminobenzaldehyde and aminoquinoline azomethine derivatives. *Russ. Chem. Bull., 63,* 1753–1758.

28. Burlov, A. S.; Mal´tsev, E. I.; Vlasenko, V. G.; Dmitriev, A. V.; Lypenko, D. A.; Garnovskii, D. A.; Uraev, A. I.;Borodkin, G. S.; Metelitsa, A, V., (2014). Synthesis, structure, photo- and electroluminescence studies of bis[2-(N-tosylamino) benzylidene-4'-dimethylaminophenylaminato]zinc.*Russ. Chem.Bull.,63,* 1759–1764.

29. Burlov, A. S.; Mal'tsev, E. I.; Vlasenko, V. G.; Garnovskii, D. A.; Dmitriev, A. V.; Lypenko, D. A.; Vannikov, A. V.; Dorovatovskii, P. V.; Lazarenko, V. A.; Zubavichus, Y. V.; Khrustalev, V. N., (2017). Synthesis, structure, photo- and electroluminescent properties of bis{(4- methyl-N-[2-[(E)-2-pyridyliminomethyl] phenyl)]benzenesulfonamide} zinc(II). *Polyhedron, 133,* 231–237.

30. Cheng, G.; So,G. K.-M.; To,W.-P.; Chen, Y.; Kwok, C.-C.; Ma, C.; Guan, X.; Chang, X.; Kwok, W.-M.; Che. C.-M., (2015).Luminescent zinc(II) and copper(I) complexes for high-performance solution-processed monochromic and white organic light-emitting devices. *Chem. Sci., 6,* 4623–4635.

31. Jafari, M. R.; Janghouri, M.; Shahedi, Z., (2017). Fabrication of an Organic Light-Emitting Diode from New Host π Electron Rich Zinc Complex. *J. Electron. Mater. 2017, 46*(1), 544–551.

32. Janghouri, M., (2017). Going from green to white color electroluminescence through a nanoscale complex of Zinc (II). *Materials Science in Semiconductor Processing, 66,* 117–122.

33. Odod, A. V.; Nikonova, E. N.; Nikonov, S. Y.; Kopylova, T. N.; Kaplunov, M. G.; Krasnikova, S. S.; Nikitenko, S. L.; Yakushchenko, I. K., (2017). *Electroluminescence Of Zinc Complexes In Various OLED Structures, Russ. Phys. J., 60*(1); DOI:10.1007/s11182-017-1038-2.

34. Li, Y.; Gao, X.; Wang L.; Tu, G., (2017). Deep-red organic light-emitting diodes with stable electroluminescent spectra based on zinc complex host material. *RSC Adv.,7,* 40533–40538.

35. Zhao, J.; Dang, F.; Liu, B.; Wu, Y.; Yang, X.; Zhou, G.; Wu, Z.; Wong, W. Y., (2017). Bis-ZnIIsalphen complexes bearing pyridyl functionalized ligands for efficient organic light-emitting diodes (OLEDs). *Dalton Trans., 46,* 6098–6110.

36. Cibian, M.;Shahalizad, A.; Souissi, F.; Castro, J.; Ferreira, J. G.; Chartrand, D.; Nunzi, J. M.; Hanan, G. S., (2018). A Zinc(II) Benzamidinate N-Oxide Complex as an Aggregation- Induced Emission Material: toward Solution-Processable White Organic Light-Emitting Devices. *Eur. J. Inorg. Chem., 39,* 4322–4330.

CHAPTER 14

Perspectives on Polymer Materials in Products Manufacturing for Green Electronics

ANDREEA IRINA BARZIC, LUMINITA IOANA BURUIANA, and
RALUCA MARINICA ALBU

*"Petru Poni" Institute of Macromolecular Chemistry,
41A Grigore Ghica Voda Alley, 700487, Iasi, Romania,
E-mail: irina_cosutchi@yahoo.com (A. I. Barzic)*

ABSTRACT

The need for green materials in the electronic industry is vital for keeping a clean environment. Given the outstanding properties of polymers, they are widely used in this domain. In this context, the chapter describes the effects of inadequate disposal of electronic wastes on the human health and the entire environment. The most important categories of green polymer materials are presented along with their latest applications. The performance of electronic components based on polymer dielectrics, semiconductors, and conductors are reviewed. The impact of green polymer systems for the bio-electronic industry is analyzed. The manner in which biodegradable and biocompatible polymer systems will contribute to the development of a green electronic market is discussed.

14.1 CURRENT STATUS AND PERSPECTIVES ON ENVIRONMENTAL IMPACT OF WASTES FROM ELECTRONICS

In the past 60 years, the electronic industry has known an outstanding evolution, which has a huge impact on everyday life. The appearance on the market of new products stimulates the consumers to buy new ones, leaving behind the "old" and technologically outdated ones. Electronic waste, (also known as "e-waste") represents a worldwide problem that negatively affects the environment as a result of piling up a large amount of electrical devices, such as washing machines, refrigerators, computers, mobile phones, digital music recorders, electronic games, microwave ovens, smartwatches, televisions, etc. [1]. Another category of e-wastes is represented by housework items, namely out of use electronics, which after that they are brought in recycling centers where they can be reused, resale, or salvaged [1]. Therefore, it could be stated that even if electronic products make our life easier, they contain over 1000 different toxic compounds, like ferrous and non-ferrous metals, polymers, ceramics, which are irremediably polluting the planet [2]. Among these, half is represented by iron and steel; almost a quarter is given by plastics, followed in a smaller percent by ferrous metals [3]. Other e-waste materials that generate undesired effects, due to their toxicity and non-biodegradability, are mercury, cadmium, lead, selenium, and hexavalent chromium and flame retardants [4, 5].

In this context, the electronics industry has a significant impact on both human health and the environment. The inadequate discharge of the end-of-life electronics will produce unwanted and harmful consequences in the environmental balance because of the presence of corrosive, inflammable, highly reactive, or radioactive elements is these products. It is well known that landfill and incineration represent the main procedures for handling the e-wastes [6]. The circular economic system is focused on the recovery of useful materials and resources from damaged electronic products that could be accomplished through an urban mining system to attain a sustainable management of materials [7–9]. Resource extraction by means of mining creates noxious consequences on the groundwater and the atmosphere. Furthermore, the mining industry can have negative social impacts [10]. The dangerous chemicals from the electronic products enter the surrounding environment. Consequently, the workers are exposed to these risky conditions and start to develop serious health problems [11]. While the Basel Convention established an important plan concerning the fighting international dumping, it's still practiced at large scale and e-wastes are still a huge problem around the world [12]. The electronics market is considered as being one of the most profitable industries and for this reason is not going to reduce its production in the future. Since the environmental burdens resulted, from this kind of wastes are mandatory to raise the industry profits, and the most consumers are unaware of the damage on the environment, it is hard to foresee if the established decontamination solutions will be applied at global level.

The improper management of the e-wastes has a strong impact on air, water, soil, and, in the end on the whole environment [13]. Burning of the throw-out electronics is not a good solution since air pollution with hydrocarbons would damage the beings which come in contact with the processing site. The generated contaminants create disastrous effects on groundwater and agriculture products, then entering in the food chain of animals and humans. Placing the wastes underwater will affect the development of the aquatic animal's plants, and microorganisms. The fauna exposed to these poisonous compounds develops severe neurological issues, infertility, breathing difficulties, skin disorders, and eventually death [2]. The soils from the zones with e-wastes are contaminated by heavy metals and organic compounds. Literature reports [14–21] indicate contamination with polychlorinated biphenyls in soils and vegetation, along with the presence of several heavy metals (Sn, Sb, Cd, Pb, Cu, Zn, Cr, Hg, Co, Se, As) and variable amounts of brominated dioxins. Inappropriately processing or storing the e-waste led to a high level of contamination with Cd and Cu in soils near vegetable areas, whereas Cd and Pb were found in the edible tissues of vegetables. Pollutant emission and novice elements appear in dockyards and in management waste areas, too. The presence of biphenyls and dibenzofurans causes a large possibility of producing tumors. The heavy metals from soils arrive in groundwater and in native plants, generating the acidification of the water, which has undesired effects on the environment.

The pollution generated by e-waste actions causes genotoxic effects on the human body, destroying the health of the regional habitats and the future generations remaining in the same environment [22–26]. For instance, high levels of Pb produce urinary and reproductive issues, or severe anemia, while Hg injures peripheral nervous systems and creates tubular dysfunction in the urinary system. When inorganic Hg is found in water, it became a pesticide (methylated mercury) and then it is accumulated in living organisms and therefore in humans food. There are several studies that reveal how the presence of cadmium and beryllium in the environment produce carcinogenic effects [25, 26]. The most seized health problems consisted of skin problems, stomach, and respiratory issues. The humans that activate in the regions with polluted soil and water tend to develop the following diseases: tuberculosis, kidneys, leukemia, respiratory system dysfunction, lung cancer, and nervous disorders [22]. Unfortunately, children are the more sensible to such issues and they need to avoid exposure to e-wastes. Because of the unsafe recycling methods performed by their family or by the children themselves, most of them are affected by noxious chemicals derived from e-wastes. Other circumstances that may contribute to such issues are dumpsites located close to the children's' homes, schools, and play areas [23, 24]. In the growing and evolution period, children present larger risks of hazardous chemical absorption in comparison with human adults. For example, dioxin compounds lead to developmental defects or cancer. Moreover, their immune, reproductive, nervous, and digestive systems are irreversibly affected when young humans come in contact with harmful substances deteriorating their further evolution [22].

Green electronics implies a cross-disciplinary science that points out the principles and applications of green technologies and engineering [27]. To avoid the long-term damaging effects on the environment, an effective integrated waste disposal system should be proposed and respected. The basic principles of this system should rely on reduction, reutilization, recycling, recovery of energy, and redesign of the material [28, 29]. In this context, the existing legal frameworks and the technologies are entirely suitable for effectively handling e-wastes based on nanomaterials (NMs) [30]. So, this urgent issue requires identification of a safe, secure, and sustainable approach to recycling. It is well known that NMs do not behave in the same way as the solid wastes, thus exhibiting negative impact on the environment and human health. Since nanoparticles (NPs) can enter the human body by dermal contact, ingestion, or, inhalation, so the potential side effects on human health are not yet fully clarified [31]. In one report, it was proved that the carbon nanotubes (CNTs) may cause cell degeneration, necrosis, and apoptosis in macrophage cell lines [32]. Therefore, there is an urgent matter to develop research programs, which would provide solutions to avoid nanotoxicology of these e-wastes.

New energy technology is in high claim since it can provide charging in tens of minutes instead of hours, or store electricity and release it to the grid at costs comparable to those of natural-gas generation. In the next years, green electronics must introduce innovative notions concerning the molecular charge transport (CT) or storage [33]. Based on interdisciplinary studies, this domain should be able to reach new goals regarding biodegradable devices or integration of bioelectronics with tissues [34]. All these should be accomplished by reducing to minimum the environmental effects, so many materials and technologies should be adapted to this purpose. The new green technologies should be focused on enhancing the device efficiency, lowering its cost, while avoids the production of secondary pollutants and proposes adequate strategies for the recovery of energy and resources.

Non-governmental organizations developed some rules for the manufacturers to guide them towards "green" materials. Some steps are made in that direction, including [35]:

- Devices that exclude halogen-derived components, but their fabrication is more environmentally expensive,
- Substitution of bromide combustion retarders with other substances more compatible with the environment,
- Application of new laws that restrict utilization of toxic compounds.

As a general conclusion, this industry should not ignore the positive environmental effects resulted from the management, collection, transportation, and the materials recovery in order to preserve a healthy planet for the future of the humanity.

14.2 MAIN TYPES OF "GREEN" POLYMER MATERIALS FOR ELECTRONICS

As the technologies used in electronics are evolving, they begin to demand new materials with increasing performance to fulfill multiple resistance and functionality criteria. When discussing "green" polymer materials involved in electronic components manufacturing, there are several kinds of macromolecular compounds that have found applicability in this area, as wearable consumer electronics, stretchable interconnects, degradable circuit boards or complex electronic skin [36]. Other investigations were directed towards biocompatible electronics leading to considerable progress concerning toward epidermal [37] and implantable devices.

Based on the nature examples, scientists were inspired to produce biodegradable forms of electronics, reaching up to transient electronics, which is a relatively new trend that has the advantage that it self-destroy after service life [38]. Literature presents some reviews regarding the main types materials [34, 36], but here the importance of polymers in this research domain is emphasized. Based on the provenience source, these materials can be divided as follows: natural polymer materials, biomacromolecules, and synthetic polymer systems. Figure 14.1 displays a basic classification of the most relevant categories of "green" polymer materials with importance in electronics.

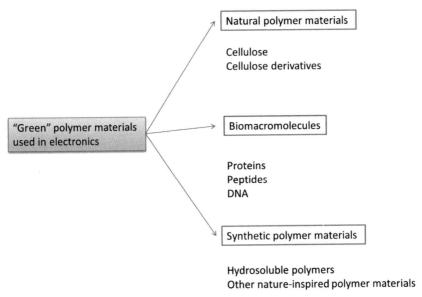

FIGURE 14.1   The classification of the main categories of "green" polymer materials used in electronics.

Table 14.1 lists some basic features of the main categories of "green" polymer materials, along with their applications based on the latest developments in electronics and bio-electronics [39–76].

14.2.1   *CELLULOSE AND CELLULOSE DERIVATIVES*

The dielectric and mechanical features, of cellulose combined with its biodegradability, makes this natural polymer attractive for manufacturing components for electronic devices that are not harmful for the environment. In addition, the biocompatibility of cellulose allows expansion of its utility towards bioelectronics. The major inconvenient arises from its high

TABLE 14.1 The Main Categories of "Green" Polymer Materials, the Degradation Temperature (T_d), Young Modulus (E), and Main Applications in Electronics and Bio-Electronics

"Green" Polymer Material Category	"Green" Features and Applications			References
	T_d, °C	**E, GPa**	**Uses**	
Cellulose and its Derivatives				
Cellulose	266	25	Electrodes, nanofibril supports	[39–42]
Cellulose nitrate	212	0.005	Printing inks	[43–45]
Cellulose acetate	330	1.2	Bioplastics	[46–48]
Hydroxypropyl cellulose		0.70	Thermo-responsive aqueous foams	[49–52]
Methylcellulose	308		Electrolyte ion gels	[53, 54]
Ethylcellulose	190	0.86	Conductive pastes and inks	[55–57]
Biomacromolecules				
DNA	>190		Sensors, biomarkers	[58–60]
Proteins	>50		Nanowires, capacitors	[61, 62]
Peptides			Field-effect transistors	
Hydrosoluble Polymers				
Poly(vinyl alcohol)	239	1.50	Alignment layers	[63–66]
Polyvinylpyrrolidone	390	1.10	Separator for ion batteries	[67–69]
Chitosan	326	0.008	Switching memory	[70–72]
Polyamide	212	1.93	Connectors, printed circuit boards	[73–75]
Other Nature-Inspired Polymer Materials				
Starch-chitosan materials			Wearable green electronics	[76]
Poly(vinyl alcohol)/Chitosan composites			Materials with programmable biodegradation	[77]

molecular weight and its crystalline structure, which affects the solubility of cellulose. For this reason, chemical modification of cellulose seems to be the best solution to impart solubility and enhance processability.

Petritz et al have doped tri-methyl silyl cellulose and have analyzed its performance as ultrathin dielectric layer for inverter production [78]. They proved that their hybrid and biodegradable gate dielectric derived from a cellulose derivative is suitable for low-voltage complementary electric circuits with outstanding high noise margins.

Liu and collaborators [79] have prepared conductive cellulose materials by in situ polymerization of aniline monomers in the cellulose matrix. At a percent of 24.6% of aniline, the conductivity of the cellulosic material reaches 0.06 $S \cdot cm^{-1}$. From this point of view, the material is useful for fabrication of supercapacitor-based energy storage components.

Cellulose-based ionogels represent a good alternative for preparation of high capacitance gate dielectric components of transistors [80]. The gate dielectric is composed of a thin flexible electrolyte layer with ionic conducting character, however strongly electronically insulating. The resulting cellulose ionogel displays a high specific capacitance ranging in the interval of 4.6–15.6 μF cm^{-2}.

Cellulose/chitosan blends lead to bio-based thin films (TFs), which upon doping with magnetite and glycerol gain better physico-chemical properties [81]. The addition of glycerol determines higher flexibility, wettability, and lower barrier characteristics, while magnetite enhances the electrical properties. The heterogeneous films have a thermal stability up to 150°C. The combination of these two fillers leads to materials with higher charge storage and insulation properties. The biodegradable films have applicability as insulators in capacitors, with diminished toxicity and potential for renewable wastes.

Electronic systems manufactured from renewable and biodegradable cellulosic materials were also reported by Jung et al [42]. Their work showed that flexible cellulose nanofibril papers have a great potential in flexible microwave and digital electronics. Besides their greater flexibility in comparison with the performance of their rigid counterparts, the electrical components made of cellulose are susceptible to fungal biodegradation, highlighting their feasibility in manufacturing flexible electronics with friendly impact on the environment. Therefore, apart from its good dielectric features, cellulose is highly promising as substrate for organic electronics. Gomez and Steckl [82] revealed that the transparency and lightweight properties of cellulose films could be successfully exploited in organic light-emitting diodes (OLEDs). In particular, nanocellulose-based paper is characterized by low roughness in comparison with regular paper. The biocompatibility of cellulosic systems makes them adequate for medical applications, which in combination with electronics will help to develop devices for diagnostic of patients [83].

14.2.2 PROTEINS AND PEPTIDES

Proteins and peptides are less used in electronic industry. Relatively recent investigations revealed that albumen extracted from chicken eggs is useful as gate dielectric for organic field transistors [84]. The films prepared from albumen were characterized by impressive dielectric properties (permittivity of 5.3–6.1) and low surface roughness (~1.5 nm). It was noticed that n- and the p-type devices show no obvious hysteresis. In a more complex study, Wu et al. [85] demonstrated that using albumen coupled layers in transistors one can imitate the synaptic activity in brain. This can be done by controlling the level of hydration of albumen dielectrics and implicitly the proton traveling in polypeptide chains subjected to the electric field. The nonvolatile memory effect was evidenced for several scan rates and ranges of the gate voltage. It is important to mention that in-plane gates together with lateral electric-double-layer help the synaptic device to simulate the spike-modulated movement of the neurotransmitters analogous to human brain. In this case, the albumen layer and its constituent ions can be viewed as the synaptic fork and neurotransmitters.

Amyloid proteins can assemble in certain conditions into nanofilaments that are poorly conductive, but simply inserting a cysteine into an amyloid monomer one may obtain wires of 100 nm diameter which exhibit electrical close to that of a metal wire [86]. Lakshmanan et al reported that there is a variety of other proteins that can form nanofilaments, which interact with metals to achieve conductive nanowires with applicability in energy storage area [87].

Electron transport proteins in lipid membranes were also studied [88–90]. Electron hopping among the proteins with electron transport can be produced from bacteria [91, 92]. Such nanowires can be prepared from *Shewanella oneidensis*, particularly from the cytochrome-rich membrane extensions. If the cytochromes are sheared from the cells, subsequently subjected to drying and chemical fixation, the electrical conductivity reach values up to 1 S cm^{-1}. When significant shrinkage of the filaments is achieved, the distance among the cytochromes is substantially reduced to almost 1 nm and in this situation the hopping mechanism of electron transport is improved [93]. The synergism between CT characteristics of the dried filaments, combined with good mechanical strength similar to that of some organic polymers [94] and transistor-like response makes these materials wonderful candidates for organic semiconducting NMs in a variety of devices, namely OLEDs, biosensors or photovoltaics (PV) [94]. The chemical fixation is done with glutaraldehyde which is toxic and this was avoided by making *S. oneidensis* wires, resulting filaments with analogous features useful for green electronics.

Other report [95] shows that the parameters of organic transistors, like threshold voltage, can be modified to enhance device performance. This can be achieved through a dielectric film based on a genetically engineered polypeptide, namely a polypeptide of a certain amino acid sequence, which preferentially interacts silicon dioxide. The adapted the organic semiconductor-dielectric interface in the transistor and controlled the threshold voltage by means of the peptide assembling conditions (particularly pH). When the peptide solutions were prepared in acidic conditions, a positive shift of the threshold voltage was noticed, whereas samples with basic pH lead to the reverse situation. An improved circuit performance and reduced power consumption of this type of transistors can be attained.

14.2.3   DNA

The DNA is a biomacromolecule with long fibrous aspect, consisting of many repeated units of deoxyribose and a phosphate unit alternately linked to nucleobase molecules. In the past years, DNA has attracted a great deal of interest in organic electronics making connections between functional electronics and organisms. The solubility of DNA is limited in most solvents, except for aqueous media. Therefore, the utilization of standard processing methodologies in organic solvents is quite difficult. DNA materials have also been used in transistors, optical waveguides, and OLEDs [96–100].

Yumusak et al. [101] prepared a DNA-lipid complex with limited solubility in water and improved solubility in organic liquids. The DNA-based films obtained by spin coating were tested as gate dielectric in transistors with lower operational voltages and reduced hysteresis as a function of degree of cross-linking of the DNA complex. It was demonstrated that DNA capping coatings augment charge injection in organic transistors since it acts as interlayer for ambipolar semiconductor allowing the injection of both kinds of charge carriers [102]. Based on this, the field-effect mobility is raised up to one order of magnitude in regards with the samples that do not contain DNA interlayer.

On the other hand, the optical waveguide is fabricated using DNA core as a multi-slab structure [98]. The DNA layer displays various electrical characteristics and might be viewed as a reconfigurable material, which is further introduced between two optical metal sheets. In this way, the current and voltage values can be closely controlled and the divergence of the DNA sorts can be considered as an optical switch. When placing the DNA between silver layers or gold/silver the performance is higher, resulting switchable features of the devices which were not reported for classical plasmonic waveguides.

From OLEDs perspective, in the first stage DNA was analyzed as host for lumophore, but the results were not as expected [103]. Later, DNA-based materials were tested as electron blocking/hole transporting layer [103], generating higher efficiency, and luminescence. Introduction of DNA in fluorescent compounds induced higher brightness of OLEDs with one order of magnitude. Another approach relies on utilization of aromatic surfactant in DNA preparation for blue OLED devices [104]. The major advantage is lower operation voltage, maximum luminance, and luminous efficacy in regard with the device lacking DNA. There is a disadvantage of DNA arising from its extensive processing to obtain films [105]. Addition of NPs in DNA layers leads to better OLED performance [104]. The idea of introduction of dye molecules in DNA matrix is useful for OLEDs due to electron resistive ability of this biopolymer [97, 106, 107]. Another innovation in this area was reported by Kobayashi and coworkers [108], which prepared active layers for OLEDs based on DNA/conducting polymer. In this context, they obtained a controlled color-tunable device.

14.2.4 HYDROSOLUBLE POLYMERS

Among the hydrosoluble polymers that have potential in "green" electronics, one can mention poly(vinyl alcohol), polyvinylpyrrolidone, chitosan, and polyamide.

Flexible and transparent poly(vinyl alcohol) films can be used to manufacture alignment layers for nematics used in flat panel displays [65, 66] or flexible substrates for circuits [109]. It was found that the anchoring strength can be significantly modified through variation of the alignment film thickness and the prepared electrically-dependent birefringence liquid crystal cells led to good device performance [65]. The biodegradable character of poly(vinyl alcohol) is suitable for transient electronics [109]. The polymer foils were processed by 3D-printing technology and then were used to manufacture CNTs-based transistors. In this way, the transience time can be controlled and selective or partial destruction of the device can be attained [109].

Another water-soluble polymer is polyvinylpyrrolidone which was used in the preparation of separator for ion batteries [69, 110]. Rechargeable transient Li-ion batteries were designed using this polymer as the separator, which enhanced the light switching capacity with a threshold voltage of about 1.6 V [110].

Hosseini and co-workers [72] used chitosan doped with Mg/Ag for the fabrication of switching memories. The device displays low power operation, the ability to retain data for a long period of time, whereas the steps for device preparation were quite simple. According to Hosseini [72], the bipolar resistive switching properties were ascribed to "trap-related space-charge-limited conduction in high resistance state and filamentary conduction in low resistance state." The resulted memory device is biocompatible and its electrodes are made of naturally abundant compounds so it can be used in transient electronics and in bio-electronics.

Rydz et al [75] revealed that polyesters, like polylactides or polyamides, prepared from renewable resources are useful for electronic applications. For instance, polyamides present chemical stability, combined with a good balance between mechanical and electrical properties, especially toughness and wear resistance. Therefore, polyamides were tested as plug and socket connectors or printed circuit boards [75].

14.2.5   OTHER NATURE-INSPIRED POLYMER MATERIALS

Sustainable and biodegradable polymer foils were prepared by blending poly(vinyl alcohol) and chitosan [77]. These materials were characterized to show their suitability as substrates for transient bioelectronics.

The material composition was proved to influence mechanical, thermo-physical characteristics and most important the transience rate. Moreover, the transience behavior of the poly(vinyl alcohol)/chitosan systems were varied through temperature. The outstanding performance and programmable degradation of these materials open new perspectives in transient biomedical and electronic devices.

The combination of chitosan with edible starch was tested by Miao et al [76] for wearable green electronic applications. They obtained biodegradable electrodes with good flexibility by incorporating 3D-interconnected conductive carbon-based fillers in starch-chitosan material. The transparent electrode presents a remarkable optoelectronic performance, namely, at 550 nm the transmittance is almost 83.5%, and the electrode sheet resistance is 46 Ω/sq. The biocompatibility of the resulted electrode can be easily adapted to skin topography or other analogous natural surfaces. At room temperature, the starch-chitosan derived systems might suffer biodegradation in lysozyme solution with the advantage that no toxic compounds are produced. The complex filler PEDOT can be recycled and further used in the synthesis of other materials. This approach [76] is promising for wearable green optoelectronics or edible electronic components.

14.3   ECO-FRIENDLY AND "GREEN" POLYMER-BASED COMPONENTS FOR ORGANIC ELECTRONICS

14.3.1   DIELECTRICS

Polymers are mainly insulators, so in this category of "green" materials, one should mainly consider those with biodegradability, which do not harm the environment after the device disposal. Such materials are mainly utilized to produce capacitors, and for this reason are essential for capacitive sensing and organic transistors. When designing a transistor, it is preferable to achieve a large capacitance per area to enable lower voltage operation. Capacitance per area ranges linearly to dielectric constant (κ) and inversely proportional to the dielectric film thickness. Thus, the polymer electrical properties and processing technique are key factors in selecting the adequate dielectric.

In order to obtain biodegradable dielectrics, the general solution is to incorporate high-κ fillers in the degradable polymer matrix [111].

Among the reinforcement agents introduced in polymers, one may mention SiO_2 (κ = 3.9), Al_2O_3 (κ = 9), AlN (κ = 9.3), HfO_2 (κ = 25), or $BaTiO_3$ (κ = 5000) [111–113]. As an example, insertion of Al_2O_3 in cellulose acetate enhanced its dielectric constant up to 27.57 at the frequency of 50 Hz [114]. Another category of fillers are those derived from carbon, such as graphene (GR), fullerene, and CNTs or nanofibers. The latter were shown to raise the value of dielectric constant of a cellulosic matrix to 3198 at 1 kHz [115]. Such multiphase dielectrics display general transience through the polymer biodegradation, while attaining tunable dielectric constants.

Plant-based fibers represent the alternative to avoid the effects of inorganic fillers. Therefore, jute, bamboo, cotton, and banana fibers are natural compounds that present excellent dielectric characteristics. Plant-based fibers are made of cellulose and lignin, which contain large amounts of free hydroxyl groups that render polarity, inducing to the composite high κ values [116]. Glucose and lactose are other natural fillers with κ values above 6 at 1 kHz, high breakdown voltages and reduced loss tangent values, i.e., 10^{-2} at 100 mHz [111].

DNA, combined with specific surfactants, was also investigated as gate dielectrics in transistors. Significant hysteresis as a result of the presence of the ion impurities and low capacitance per area was noticed [101, 111].

Poly(glycerol sebacate) is a biodegradable material with remarkable dielectric and mechanical properties. It is useful for capacitive sensors due to its capacity to withstand compression [111]. Pyramidal microstructures of poly(glycerol sebacate) were used for fabrication of capacitive pressure sensors that are eco-friendly. The insulator is sandwiched between biocompatible electrodes consisting of corrodible metals, like Mg and Fe. During thermally curing, the cross-linking of poly(glycerol sebacate) allows preparation of structured insulators. The pressure sensors have an elevated sensitivity and can identify 5 mg grain of rice, while it's in vivo degradation rate reaches 0.2–1.5 mm/month [117].

Biodegradable polymer insulators will represent the future of new sensor designs and stretchable devices. An important aspect concerning the presented information relies on the fact that the data from literature refer to low operational frequencies (<kHz). For the implementation of these materials in complex devices, it is essential to the extent of the study of dielectric performance towards the high-frequency domain.

14.3.2 POLYMER-BASED SEMICONDUCTORS

Eco-friendly semiconducting materials are vital for improving the switching mechanism of transistors, and consequently are important for complex electronic circuitry. The most studied semiconducting polymers are polythiophenes [118] and donor–acceptor copolymers created originally for organic photovoltaics (OPV), such as diketopyrrolopyrroles [119, 120].

Biodegradable semiconductors were not deeply investigated since electron-conducting polymers were mainly analyzed for recording electrical signals in the body [111]. However, such materials can help to develop more complicated biodegradable devices. Blending is a widely used method to achieve semiconducting polymer systems. The compatibility of poly(3-thiophene methyl acetate) with some biodegradable polymers (poly(tetramethylene succinate), poly(ester urea), polyurethane (PU), polylactide) was analyzed [121–123]. The resulting materials were cytocompatible, while the electric conductivity is in the range of 10^{-5}–10^{-6} S/cm [111]. Poly(3-hexylthiophene was mixed with polycaprolactone, and polylactic-co-glycolic acid and subjected to electrospinning [124, 125]. Addition of polycaprolactone reduced the charge mobility up to two orders of magnitude at a bending ratio of 50/50. The strategy of mixing rigid semiconducting polymers with inert ones is fruitful if interconnected aggregates are obtained in the semiconducting phase with locally intermolecular CT [126].

In the above-described examples, biodegradability is ensured through commonly hydrolyzable linkages. The latter can be introduced within a second component in the blend, which impedes the semiconductor from total degradation, or they can be inserted into the main chain interrupting the conjugation responsible for higher mobilities. A better approach involves the seminal use of imine bonds as conjugated linkages among diketopyrrolopyrrole and p-phenylenediamine [111]. The semiconducting material displays high hole mobility and good biodegradability within 30 days.

14.3.3 POLYMERIC CONDUCTING MATERIALS

In the category of conductive and biodegradable polymers, the most efficient are the conjugated structures which are doped into a conducting state. Such materials have properties useful for interconnects and contacts in electronic devices. Biodegradable conductors make the connection among various components of the circuit, but they also can be exploited in linking electrically responsive entities from living organisms, like neurons or cardiac cells. Doping is mandatory sometimes to attain polymers with good conducting features and if biocompatibility of the processed materials is, also desired additional considerations must be made. The most known conjugated polymers are poly(3,4-ethylenedioxythiophene) (PEDOT), polypyrrole (PPy), polyaniline (PANI), and the blend of poly(styrenesulfonate) (PSS) with PEDOT [111].

As a general classification, these materials are divided as follows:

- **Type I Conductors:** These are represented by composites of non-degradable conjugated polymers with degradable dielectric macromolecular compounds.
- **Type II Conductors:** These are prepared by combining through cleavable linkages conducting oligomers with degradable polymer.

Regardless the type of conductor, the presence of dopants is crucial to attain the desired conductivity values, and these molecules must also have good biocompatibility. Type II conducting materials do not display conductivities that overcome those of their type I counterparts.

In order to achieve biocompatibility one may insert in the conducting matrix a bio-component resulting systems that are able to biodegrade. Because the electronic counterpart is non-degradable, the goal is raise the electrical conductivity, while reducing the relative amount of the non-degradable conjugated component. Conductive polymer NPs can be added in a biodegradable matrix and the doping can be done orthogonally [127–129]. The dopant concentration must exceed the percolation threshold to ensure formation of conduction pathways inside the composite, but it must be kept in such a proportion that would not affect the biodegradable properties. The strategy is adequate for applications that are not involving elevated conduction values, namely stimulation of cells development or tissue regeneration [130]. Among the studied systems, it is worth mentioning PPy NPs within poly(D,L-lactic acid) with conductivity of 1×10^{-3} S/cm [127], PEDOT NPs containing hyaluronic acid doped introduced in poly(L-lactic acid) with conductivity of 4.7×10^{-3} S/cm [129], PANI/camphorsulfonic acid system was cospun with gelatin to achieve fibrous sheets with conductivity of 2.1×10^{-2} S/cm [131], camphorsulfonic/PANI/poly(L-lactide-co-ε-caprolactone) fibers with conductivity of 1.38×10^{-2} S/cm [132].

In order to enhance the mechanical performance of conductive polymers, flexible but non-conjugated linkers are introduced in the main chain [111]. The CT properties are still in the desired limits. When using biodegradable conjugated oligomers one may achieve products with biocompatible degradation because they have suitable sizes to allow phagocytization by macrophages that naturally travel to implantation zones as a response of immune system. After iodine doping of pyrrole-thiophene-pyrrole trimers and aliphatic sequences linked with ester bonds, the conductivity reaches a value of 10^{-4} S/cm, which is similar with that of PANI grafted to gelatin and doped with camphorsulfonic acid [111]. These conducting materials are important for recording and stimulating small bio-electronic signals. A higher control over the material chemistry and composition, will contribute to close the gap and achieve elevated the conductivities as demanded for high-performance electronic devices.

14.4 "GREEN" POLYMER MATERIALS FOR BIO-ELECTRONIC INDUSTRY

14.4.1 SUBSTRATES AND ENCAPSULATES

The bio-features of the polymer materials are excellent for testing in bio-electronic devices. The replacement of glass with transparent polymers is a good alternative for enhancing the mechanical resistance to strain/stress of the device substrate and permeation barrier properties that enable operational stability. The resistance of the device during flexing/bending/stretching is improved if the substrate thickness is reduced and by using suitable interlayers and interconnects [34]. In the same time, accomplishment of an appropriate barrier to oxygen and moisture is difficult. Alternative solutions are based on the combination of multilayer structures, where each material provides the proper barrier to one species. Synthetic polymer substrates with biocompatible characteristics are key factors in integration of electronics with living tissue [34].

As a result of their inertness, low density polyethylene (LDPE) and polydimethylsiloxane (PDMS) have become candidates for construction of devices used for in vivo and in vitro tests. PDMS was shown to have good biocompatibility, hemocompatibility, and inflammatory resistance, so it was exploited as synthetic substrate for the manufacturing of biocompatible electronic platforms that evaluate the bladder nerve function [133].

Moreover, other synthetic materials are currently utilized in the production of implantable devices, such as polyethylene glycol (PEG), parylene, polylactic-co-glycolic acid (PLGA), poly(2-hydroxyethyl methacrylate) (pHEMA), polylactic acid (PLA), PU, and polyvinyl alcohol (PVA) [34]. Among them, PVA has the advantage of water solubility so it involves green processing routes. Therefore, employment of this polymer as a coating layer for implantable devices enables the controlled transport of the water soluble analytes through diminishment of the crosslinking density and chain swelling. PVA was used as sacrificial substrate to design devices for physiological evaluation and stimulation that analyze interfaces to the human epidermis [37, 134].

14.4.2 CONTACT ELECTRODES AND INTERCONNECTS

In the past decade, a great attention has been paid to development of electrodes capable to connect electronics with biological media for the purpose of recognizing or modulation events. The latest achievement in this domain allowed controlling both types of events. This opens a novel perspective in interfacing electronics with the neuron, which could evolve towards brain/machine interfaces and artificial vision [135]. Compared to metal electrodes, interfacing electronics with biological matter through soft, conducting polymers brings the possibility to tune the transport neurotransmitters in their ionic form allowing recognition and modulation of the event. PEDOT:PSS has been prepared directly into the living neural tissue of a lab mouse [136]. The polymer filaments developed in the interstitial and extracellular spaces [136]. The filaments are entangled around and between cells and maintain their viability. This approach was further continued by Khodagholy et al [137] which used photolithography to produce microelectrodes based on PEDOT:PSS on a biocompatible and very flexible sacrificial parylene support. These materials were transplanted in the brain of a mouse to analyze from electrocorticography data the waves mimicking epileptic spikes. The PEDOT:PSS conformability to the brain surface overcome the performance of gold electrodes.

To fabricate implantable devices it is important to use materials with an intermediate functionality [34]. One such element is represented by the interconnect layer. Wong et al [138] obtained conductive bio-adhesives from PU. The reported materials display higher performance in regard to the conventional oil-based counterparts. The electrical conductivity of the bio-adhesives was tuned by the introduction of silver NPs. The shear flow behavior was adjustable relative to the water percent, making the systems compatible with many processing techniques used in applications like electrical interconnects and printed circuits.

14.5 CONCLUSIONS AND FUTURE PERSPECTIVES

The electronic industry has a great future ahead, but this should not affect the environment. Huge amounts of e-wastes should be carefully deposited and processed by means of non-polluting approaches. Up to this point, scientists introduced several types of green materials, including polymeric ones. Depending on the function that must be fulfilled in the circuit, a variety of biodegradable dielectrics, semiconductors or conductors were developed. The demand for new green polymer materials is growing and necessary for the fabrication of electrodes, interconnects, and substrates in eco-friendly electronic devices. Future advances in the synthesis of intrinsically stretchable/elastic, highly conductive biodegradable polymeric systems will significantly contribute to the creation of electronics interfaced with complex surfaces, like human skin.

KEYWORDS

- eco-friendly devices
- electronic waste
- genotoxicity
- green polymer materials
- microorganisms
- polymers

REFERENCES

1. Sitaramaiah, Y., & Kumari, M. K., (2014). Electronic waste leading to environmental pollution. *Journal of Chemical and Pharmaceutical Sciences, National Seminar on*

Impact of Toxic Metals, Minerals and Solvents Leading to Environmental Pollution,
39–42.

2. Needhidasan, S., Samuel, M., & Chidambaram, R., (2014). Electronic waste: An emerging threat to the environment of urban India. *J. Environ. Health Sci. Eng.,* 12–36.

3. Hima, B. G. N., (2014). Impact of toxic metals, minerals, solvents, e-waste, and plastics leading to environmental pollution. *Journal of Chemical and Pharmaceutical Sciences, National Seminar on Impact of Toxic Metals, Minerals and Solvents Leading to Environmental Pollution,* 124–131.

4. Widmer, R., Oswald-Krapf, H., Sinha-Khetriwal, D., Schnellmann, M., & Böni, H., (2005). Global perspectives on e-waste Environ. *Impact Assess. Rev.,* 25(5), 436–458.

5. Chatuverdi, A., Aora, R., & Killguss, U., (2012). E-waste recycling in India. In: Mukherjee, S., & Chakraborty, D., (eds.), *Environmental Scenario, In India: Successes and Predicament.* Routhledge Taylor and Francis: New York.

6. Williams, J. A. S., (2006). A review of electronics demanufacturing processes. *Resour. Conserv. Recycl.,* 47, 195–208.

7. Hertwich, E. G., Gibon, T., Bouman, E. A., Arvesen, A., Suh, S., Heath, G. A., Bergesen, J. D., Ramirez, A., Vega, M. I., & Shi, L., (2015). Integrated life-cycle assessment of electricity-supply scenarios confirms global environmental benefit of low-carbon technologies. *Proc. Natl. Acad. Sci. U.S.A.,* 112, 6277–6282.

8. Chen, P. C., Liu, K. H., & Ma, H. W., (2017). Resource and waste-stream modeling and visualization as decision support tools for sustainable materials management. *J. Cleaner Prod.,* 150, 16–25.

9. Arora, R., Paterok, K., Banerjee, A., & Saluja, M. S., (2017). Potential and relevance of urban mining in the context of sustainable cities. *IIMB Management Review,* 29, 210–224.

10. Schmidt, C. W., (2006). Unfair trade e-waste in Africa environs. *Health Perspect,* 114(4), A232–A235.

11. Ohajinwa, C. M., Van Bodegom, P. M., Vijver, M. G., & Peijnenburg, W. J. G. M., (2017). Health risks awareness of electronic waste workers in the informal sector in Nigeria. *Int. J. Environ. Res. Public Health,* 14(8), 911.

12. Mishra, S., Shamanna, B. R., & Kannan, S., (2017). Exploring the awareness regarding e-waste and its health hazards among the informal handlers in Musheerabad area of Hyderabad. *Indian J. Occup. Environ. Med.,* 21(3), 143–148.

13. *Impacts of E-Waste on the Environment.* http://www.eterra.com.ng/articles/impacts-e-waste-environment/ (accessed on 22 February 2020).

14. Wang, Y., Luo, C., Li, J., Yin, H., Li, X., & Zhang, G., (2011). Characterization of PBDEs in soils and vegetations near an e-waste recycling site in South China. *Environmental Pollution,* 159, 2443–2448.

15. Quan, S. X., Yan, B., Lei, C., Yang, F., Li, N., Xiao, X. M., & Fu, J. M., (2014). Distribution of heavy metal pollution in sediments from an acid leaching site of e-waste. *Sci. Total Environ.,* 499, 349–355.

16. Luo, C., Liu, C., Wang, Y., Liu, X., Li, F., Zhang, G., & Li, X., (2011). Heavy metal contamination in soils and vegetables near an e-waste processing site South China. *Journal Hazard. Mater.,* 186, 481–490.

17. Alam, M., & Bahauddin, K. M., (2015). Electronic waste in Bangladesh: Evaluating the situation, legislation and policy and way forward with strategy and approach. *Present Environmental and Sustainable Development, 9*, 81–101.
18. Pradhan, J. K., & Kumar, S., (2014). Informal e-waste recycling: Environmental risk assessment of heavy metal contamination in Mandoli industrial area, Delhi, India. *Environ. Sci. Pollut. Res., 21*, 7913–7928.
19. Tue, N. M., Takahashi, S., Suzuki, G., Isobe, T., Viet, P. H., Kobara, Y., Seike, N., Zhang, G., Sudaryanto, A., & Tanabe, S., (2013). Contamination of indoor dust and air by polychlorinated biphenyls and brominated flame retardants and relevance of non-dietary exposure in Vietnamese informal e-waste recycling sites. *Environ. Int., 51*, 160–167.
20. Xiao, X., Hu, J., Chen, P., Chen, D., Huang, W., Peng, P., & Ren, M., (2014). Spatial and temporal variation, source profile, and formation mechanisms of PCDD/Fs in the atmosphere of an e-waste recycling area, South China. *Environ. Toxicol. Chem., 33*, 500–507.
21. Wu, Q., Leung, J. Y. S., Geng, X., Chen, S., Huang, X., L, H., Huang, Z., Zhu, L., Chen, J., & Lu, Y., (2015). Heavy metal contamination of soil and water in the vicinity of an abandoned e-waste recycling site: Implications for dissemination of heavy metals. *Sci. Total Environ., 506–507*, 217–225.
22. Grant, K., Goldizen, F. C., Sly, P. D., Brune, M. N., Neira, M., van den, B. M., & Norman, R. E., (2013). Health consequences of exposure to e-waste: A systematic review. *Lancet Glob Health, 1*, e350–e361.
23. Zhang, A., Hu, H., Sanchez, B. N., Ettinger, A. S., Park, S. K., Cantonwine, D., Schnaas, L., Wright, R. O., Lamadrid-Figueroa, H., & Tellez-Rojo, M. M., (2012). Association between prenatal lead exposure and blood pressure in children. *Environ. Health Perspect., 120*, 445–450.
24. Kippler, M., Tofail, F., Hamadani, J. D., Gardner, R. M., Grantham-McGregor, S. M., Bottai, M., & Vahter, M., (2012). Early-life cadmium exposure and child development in 5-year-old girls and boys: A cohort study in rural Bangladesh. *Environ. Health Perspect., 120*, 1462–1468.
25. Beyersmann, D., (2002). Effects of carcinogenic metals on gene expression. *Toxicol. Lett., 127*(1–3), 63–68.
26. Cooper, R. G., & Harrison, A. P., (2009). The uses and adverse effects of beryllium on health. *Indian J. Occup. Environ. Med., 13*(2), 65–76.
27. Sdrolia, E., & Zarotiadis, G., (2019). A comprehensive review for green product term: From definition to evaluation. *J. Econom. Surv., 33*(1), 150–178.
28. Diaz, L. A., & Lister, T. E., (2018). Economic evaluation of an electrochemical process for the recovery of metals from electronic waste. *Waste Manag., 74*, 384–392.
29. Gao, M., Shih, C. C., Pan, S. Y., Chueh, C. C., & Chen, W. C., (2018). Advances and challenges of green materials for electronics and energy storage applications: From design to end-of-life recovery. *J. Mater. Chem. A., 6*, 20546–20563.
30. Nichols, G. P., (2016). Exploring the need for creating a standardized approach to managing nanowaste based on similar experiences from other wastes. *Environ. Sci.: Nano., 3*, 946–952.

31. Borm, P. J. A., Robbins, D., Haubold, S., Kuhlbusch, T., Fissan, H., Donaldson, K., et al., (2006). The potential risks of nanomaterials: A review carried out for ECETOC. *Part. Fiber Toxicol.*, *3*, 11.

32. Jia, G., Wang, H., Yan, L., Wang, X., Pei, R., Yan, T., Zhao, Y., & Guo, X., (2005). *Environ. Sci. Technol.*, *39*, 1378–1383.

33. Gao, M., Shih, C. C., Pan, S. Y., Chueh, C. C., & Chen, W. C., (2018). Advances and challenges of green materials for electronics and energy storage applications: From design to end-of-life recovery. *J. Mater. Chem. A.*, *6*, 20546–20563.

34. Irimia-Vladu, M., (2014). "Green" electronics: Biodegradable and biocompatible materials and devices for sustainable future. *Chem. Soc. Rev.*, *43*, 588–610.

35. Gaidajis, G., Angelakoglou, K., & Aktsoglou, D., (2010). E-waste: Environmental problems and current management. *J. Eng. Sci. Technol. Review*, *3*(1), 193–199.

36. Baumgartner, M., Coppola, M. E., Sariciftci, N. S., Glowacki, E. D., Bauer, S., & Irimia-Vladu, M., (2018). Emerging "green" materials and technologies for electronics. In: Irimia-Vladu, M., Glowacki, E. D., Sariciftci, N. S., & Bauer, S., (eds.), *Green Materials for Electronics* (pp. 1–54). Wiley: New York.

37. Kim, D. H., Lu, N., Ma, R., Kim, Y. S., Kim, R. H., Wang, S., et al., (2011). Epidermal electronics. *Science*, *333*, 838–843.

38. Kang, S. K., Murphy, R. K. J., Hwang, S. W., Lee, S. M., Harburg, D. V., Krueger, N. A., et al., (2016). Bioresorbable silicon electronic sensors for the brain. *Nature*, *530*, 71–76.

39. Nada, A. M. A., & Hassan, M. L., (2000). Thermal behavior of cellulose and some cellulose derivatives. *Polymer Degradation and Stability*, *67*, 111–115.

40. Eichhorn, S. J., & Young, R. J., (2001). The Young's modulus of microcrystalline cellulose. *Cellulose*, *8*, 197–207.

41. Liu, S., Yu, T., Wu, Y., Li, W., & Li, B., (2014). Evolution of cellulose into flexible conductive green electronics: A smart strategy to fabricate sustainable electrodes for super capacitors. *RSC Adv.*, *4*(64), 34134–34143.

42. Jung, Y. H., Chang, T. H., Zhang, H., Yao, C., Zheng, Q., Yang, V. W., et al., (2015). High-performance green flexible electronics based on biodegradable cellulose nanofibril paper. *Nature Comms.*, *6*, 7170.

43. Rong, M. R., & Li, X. G., (1998). Thermal degradation of cellulose and cellulose esters. *J. Appl. Polym. Sci.*, *68*, 293–304.

44. Fallah, F., Khorasani, M., & Ebrahimi, M., (2017). Improving the mechanical properties of waterborne nitrocellulose coating using nano-silica particles. *Prog. Org. Coats.*, *109*, 110–116.

45. Ash, M., & Ash, I., (2004). *Handbook of Green Chemicals* (p. 645). Synapse Information Resources: USA.

46. Gaan, S., Mauclaire, L., Rupper, P., Salimova, V., Tran, T. T., & Heuberger, M., (2011). Thermal degradation of cellulose acetate in presence of bis-phosphoramidates. *J. Anal. Appl. Pyrolysis*, *90*, 33–41.

47. Rynkowska, E., Fatyeyeva, K., Kujawa, J., Dzieszkowski, K., Wolan, A., & Kujawski, W., (2018). The effect of reactive ionic liquid or plasticizer incorporation on the physicochemical and transport properties of cellulose acetate propionate-based membranes. *Polymers*, *10*(86), 1–18.

48. Park, H. M., Misra, M., Drzal, L. T., & Mohanty, A. K., (2004). "Green" nanocomposites from cellulose acetate bioplastic and clay: Effect of eco-friendly triethyl citrate plasticizer. *Biomacromolecules, 5*(6), 2281–2288.

49. Yanagida, N., & Matsuo, M., (1992). Morphology and mechanical properties of hydroxypropyl cellulose cast films cross linked in solution. *Polymer, 33*(5), 996–1005.

50. Weißenborn, E., & Braunschweig, B., (2019). Hydroxypropyl cellulose as a green polymer for thermo-responsive aqueous foams. *Soft Matter., 15*, 2876–2883.

51. Mori, N., Morimoto, M., & Nakamura, K., (1999). Hydroxypropylcellulose films as alignment layers for liquid crystals. *Macromolecules, 32*(5), 1488–1492.

52. Cosutchi, A. I., Hulubei, C., Stoica, I., & Ioan, S., (2011). A new approach for patterning epiclon-based polyimide precursor films using a lyotropic liquid crystal template. *J. Polym. Res., 18*, 2389–2402.

53. Li, X. G., Huang, M. R., & Bai, H., (1999). Thermal decomposition of cellulose ethers. *J. Appl. Polym. Sci., 73*, 2927–2936.

54. Pérez-Madrigal, M. M., Edo, M. G., & Alemán, C., (2016). Powering the future: Application of cellulose-based materials for supercapacitors. *Green Chem., 18*(22), 5930–5956.

55. Lai, H. L., Pitt, K., & Craig, D. Q., (2010). Characterization of the thermal properties of ethylcellulose using differential scanning and quasi-isothermal calorimetric approaches. *Int. J. Pharm., 386*(1/2), 178–184.

56. Kangarlou, S., & Haririan, I., (2007). Physico-mechanical analysis of free ethylcellulose films plasticized with incremental weight percents of dibutyl sebacat. *Iran. J. Pharm. Sci., 3*(3), 135–142.

57. Antunes, B. R. A., (2015). Cellulose-based composites as functional conductive materials for printed electronics. PhD Dissertation. Faculty of Sciences and Technology of Nova University of Lisbon, Lisbon.

58. Karni, M., Zidon, D., Polak, P., Zalevsky, Z., & Shefi, O., (2013). Thermal degradation of DNA. *DNA Cell Biol., 32*(6), 298–301.

59. Wang, J., (2000). From DNA biosensors to gene chips. *Nucleic Acids Res., 28*(16), 3011–3016.

60. Ziegler, A., Koch, A., Krockenberger, K., & Großhennig, A., (2012). Personalized medicine using DNA biomarkers: A review. *Hum. Genet., 131*(10), 1627–1638.

61. Bischof, J. C., & He, X., (2005). Thermal stability of proteins. *Ann. N.Y. Acad. Sci., 1066*, 12–33.

62. Lovley, D. R., (2017). E-biologics: Fabrication of sustainable electronics with "green" biological materials. *Mbio., 8*(3), e00695–17.

63. Magalhães, M. S., Filho, R. D. T., & Fairbairn, E. M. R., (2013). Durability under thermal loads of polyvinyl alcohol fibers. *Matéria (Rio J.), 18*, 1587–1595.

64. Yamaura, K., Tada, M., Tanigami, T., & Matsuzawa, S., (1986). Mechanical properties of films of poly(vinyl alcohol) derived from vinyl trifluoroacetate. *J. Appl. Polym. Sci., 31*, 493–500.

65. Cui, Y., Zola, R. S., Yang, Y. C., & Yang, D. K., (2012). Alignment layers with variable anchoring strengths from polyvinyl alcohol. *J. Appl. Phys., 111*, 063520.

66. Nechifor, C. D., Postolache, M., Albu, R. M., Barzic, A. I., & Dorohoi, D. O., (2019). Induced birefringence of rubbed and stretched polyvinyl alcohol foils as alignment layers for nematic molecules. *Polym. Adv. Technol.*

67. Loría-Bastarrachea, M. I., Herrera-Kao, W., Cauich-Rodríguez, J. V., Cervantes-Uc. J. M., Vázquez-Torres, H., & Avila-Ortega, A., (2011). A TG/FTIR study on the thermal degradation of poly(vinyl pyrrolidone). *J. Therm. Anal. Calorim., 104,* 737–742.

68. Bernal, A., Kuritka, I., & Saha, P., (2011). Poly(vinyl alcohol)-poly(vinyl pyrrolidone) blends: Preparation and characterization for a prospective medical application. *Mathematical Methods and Techniques in Engineering and Environmental Science,* pp. 431–434.

69. Landi, G., Neitzert, H. C., & Sorrentino, A., (1990). New biodegradable nano-composites for transient electronics devices. *AIP Conference Proceedings, 020012* (2018).

70. Hong, P. Z., Li, S. D., Ou, C. Y., Li, C. P., Yang, L., & Zhang, C. H., (2007). Thermo-gravimetric analysis of chitosan. *J. Appl. Polym. Sci., 105,* 547–551.

71. Le, H. R., Qu, S., Mackay, R. E., & Rothwell, R., (2012). Fabrication and mechanical properties of chitosan composite membrane containing hydroxyapatite particles. *J. Adv. Ceram., 1*(1), 66–71.

72. Hosseini, N. R., & Lee, J. S., (2015). Biocompatible and flexible chitosan-based resistive switching memory with magnesium electrodes. *Adv. Funct. Mater., 25,* 5586–5592.

73. He, Y., Chen, S., Zheng, Q., & Chen, Y., (2014). Thermal stability and yellowing of polyamide finished with a compound anti-thermal-yellowing agent. *J. Tex. I., 106*(12), 1263–1269.

74. http://www.designerdata.nl/plastics/thermo+plastics/PA12, (accessed on 22 February 2020).

75. Rydz, J., Sikorska, W., Kyulavska, M., & Christova, D., (2015). Polyester-based (Bio) degradable polymers as environmentally friendly materials for sustainable development. *Int. J. Mol. Sci., 16,* 564–596.

76. Miao, J., Liu, H., Li, Y., & Zhang, X., (2018). Biodegradable transparent substrate based on edible starch-chitosan embedded with nature-inspired three-dimensionally interconnected conductive nanocomposites for wearable green electronics. *ACS Appl. Mater. Interfaces, 10*(27), 23037–23047.

77. Liu, L., Liang, H., Zhang, J., Zhang, P., Xu, Q., Lu, Q., & Zhang, C., (2018). Poly(vinyl alcohol)/chitosan composites: Physically transient materials for sustainable and transient bioelectronics. *J. Clean. Prod., 195,* 786–795.

78. Petritz, A., Wolfberger, A., Fian, A., Griesser, T., Irimia-Vladu, M., & Stadlober, B., (2015). Cellulose-derivative-based gate dielectric for high performance organic complementary inverters. *Adv. Mater., 27,* 7645–7656.

79. Liu, S., Yu, T., Wu, Y., Li, W., & Li, B., (2014). Evolution of cellulose into flexible conductive green electronics: A smart strategy to fabricate sustainable electrodes for super capacitors. *RSC Adv., 4,* 34134–34143.

80. Setti, L., Fraleoni-Morgera, A., Mencarelli, I., Filippini, A., Ballarin, B., & Di Biase, M., (2007). An HRP-based amperometric biosensor fabricated by thermal inkjet printing. *Sens. Actuators, B., 126*(1), 252–257.

81. Arantes, A. C. C., Silva, L. E., Wood, D. F., Almeida, C. D. G., Tonoli, G. H. D., Oliveira, J. E., Silva, J. P. D., Williams, T. G., Orts, W. J., & Bianchi, M. L., (2019). Bio-based thin films of cellulose nanofibrils and magnetite for potential application in green electronics. *Carbohydr. Polym., 207,* 100–107.

82. Gomez, E. F., & Steckl, A. J., (2015). Improved performance of OLEDs on cellulose/ epoxy substrate using adenine as a hole injection layer. *ACS Photonics.*, *2*(3), 439–445.
83. Klemm, D., Kramer, F., Moritz, S., Lindström, T., Ankerfors, M., Gray, D., & Dorris, A., (2011). Nanocelluloses: A new family of nature-based materials. *Angew. Chem. Int. Ed.*, *50*(24), 5438–5466.
84. Chang, J. W., Wang, C. G., Huang, C. Y., Da Tsai, T., Guo, T. F., & Wen, T. C., (2011). Chicken albumen dielectrics in organic field-effect transistors. *Adv. Mater.*, *23*(35), 4077–4081.
85. Wu, G., Feng, P., Wan, X., Zhu, L., Shi, Y., & Wan, Q., (2016). Artificial synaptic devices based on natural chicken albumen coupled electric-double-layer transistors. *Sci. Rep.*, *6*, 23578.
86. Scheibel, T., Parthasarathy, R., Sawicki, G., Lin, X. M., Jaeger, H., & Lindquist, S. L., (2003). Conducting nanowires built by controlled self-assembly of amyloid fibers and selective metal deposition. *Proc. Natl. Acad. Sci. U.S.A.*, *100*, 4527–4532.
87. Lakshmanan, A., Zhang, S., & Hauser, C. A. E., (2012). Short self-assembling peptides as building blocks for modern nanodevices. *Trends Biotechnol.*, *30*, 155–165.
88. Leung, K. M., Wanger, G., El-Naggar, M. Y., Gorby, Y., Southam, G., Lau, W. M., & Yang, J., (2013). *Shewanella oneidensis* MR-1 bacterial nanowires exhibit p-type, tunable electronic behavior. *Nano. Lett.*, *13*, 2407–2411.
89. Subramanian, P., Pribadian, S., El-Naggar, M. Y., & Jensen, G. J., (2018). Ultrastructure of Shewanella oneidensis MR-1 nanowires revealed by electron cryotomography. *Proc. Natl. Acad. Sci. U.S.A.*, *115*(14), E3246–E3255.
90. Polizzi, N. F., Skourtis, S. S., & Beratan, D. N., (2012). Physical constraints on charge transport through bacterial nanowires. *Faraday Discuss*, *155*, 43–62.
91. El-Naggar, M. Y., Gorby, Y. A., Xia, W., & Nealson, K. H., (2008). The molecular density states in bacterial nanowires. *Biophys. J.*, *95*, L10–LL12.
92. Pirbadian, S., Barchinger, S. E., Leung, K. M., Byun, H. S., Jangir, Y., Bouhenni, R. A., et al., (2014). *Shewanella oneidensis* MR-1 nanowires are outer membrane and periplasmic extensions of the extracellular electron transport components. *Proc. Natl. Acad. Sci. U.S.A.*, *111*, 12883–12888.
93. Pirbadian, S., & El-Naggar, M. Y., (2012). Multistep hopping and extracellular charge transfer in microbial redox chains. *Phys. Chem. Chem. Phys.*, *14*, 13802–13808.
94. Leung, K. M., Wanger, G., Guo, Q., Gorby, Y., Southam, G., Lau, W. M., & Yang, J., (2011). Bacterial nanowires: Conductive as silicon, soft as polymer. *Soft Matter.*, *7*, 6617–6621.
95. Dezieck, A., Acton, O., Leong, K., Oren, E. E., Ma, H., Tamerler, C., Sarikaya, M., & Jen, A. K. Y., (2010). Threshold voltage control in organic thin film transistors with dielectric layer modified by a genetically engineered polypeptide. *Appl. Phys. Lett.*, *97*(1), 1–4.
96. Steckl, A. J., (2007). DNA: A new material for photonics? *Nat. Photonics.*, *1*, 3–5.
97. Hagen, J. A., Li, W., Steckl, A. J., & Grote, J. G., (2006). Enhanced emission efficiency in organic light-emitting diodes using deoxyribonucleic acid complex as an electron blocking layer. *Appl. Phys. Lett.*, *88*(17), 10–13.
98. Ebrahimi, S., Sabbaghi-Nadooshan, R., & Tavakoli, M. B., (2018). DNA implementation for optical waveguide as a switchable transmission line and memristor. *Opt. Quant. Electron.*, *50*, 196.

99. Yang, X., Wang, Z. Y., Zhou, J., Wong, C. Y., & Pun, E. Y. B., (2009). Low loss DNA biopolymer optical waveguide. *2009 IEEE LEOS Annual Meeting Conference Proceedings, 11000505*(2009).
100. Gomez, E. F., Venkatraman, V., Grote, J. G., & Steckl, A. J., (2014). DNA Bases thymine and adenine in bio-organic light emitting diodes. *Sci. Rep., 4*(7105).
101. Yumusak, C., Singh, T. B., Sariciftci, N. S., & Grote, J. C., (2009). Bio-organic field effect transistors based on crosslinked deoxyribonucleic acid (DNA) gate dielectric. *Appl. Phys. Lett., 95,* 263304.
102. Zhang, Y., Zalar, P., Kim, C., Collins, S., Bazan, G. C., & Nguyen, T. Q., (2012). DNA interlayers enhance charge injection in organic field-effect transistors. *Adv. Mater., 24,* 4255–4260.
103. Haque, S., Sher Shah, M. S. A., Rahman, M., & Mohiuddin, M., (2017). Biopolymer composites in light emitting diodes. In: Sadasivuni, K. K., Ponnamma, D., Kim, J., Cabibihan, J. J., Deepalekshmi, & Al-Maadeed, M. A., (eds.), *Biopolymer Composites in Electronics.* Elsevier: Amsterdam.
104. Chen, I. C., Chiu, Y. W., & Hung, Y. C., (2012). Efficient biopolymer blue organic light-emitting devices with low driving voltage. *Jpn. J. Appl. Phys., 51,* 031601.
105. Heckman, E. M., Hagen, J. A., Yaney, P. P., Grote, J. G., & Hopkins, F. K., (2005). Processing techniques for deoxyribonucleic acid: Biopolymer for photonics applications. *Appl. Phys. Lett., 87,* 211115.
106. Sun, Q., Subramanyam, G., Dai, L., Check, M., Campbell, A., Naik, R., Grote, J., & Wang, Y., (2009). Highly efficient quantum-dot light-emitting diodes with DNA-CTMA as a combined hole-transporting and electron blocking layer. *ACS Nano, 3,* 737–743.
107. Nakamura, K., Ishikawa, T., Nishioka, D., Ushikubo, T., & Kobayashi, N., (2010). Color-tunable multilayer organic light emitting diode composed of DNA complex and tris(8-hydroxyquinolinato)aluminum. *Appl. Phys. Lett., 97,* 193301.
108. Kobayashi, N., (2011). Boiled with DNA/conducting polymer complex as active layer. *Nonl. Opt. Quant. Opt., 42,* 233–251.
109. Yoon, J., Han, J., Choi, B., Lee, Y., Kim, Y., Park, J., Lim, M., Kang, M. H., Kim, D. H., Kim, D. M., Kim, S., & Choi, S. J., (2018). Three-dimensional printed poly(vinyl alcohol) substrate with controlled on-demand degradation for transient electronics. *ACS Nano., 12*(6), 6006–6012.
110. Fu, K., Liu, Z., Yao, Y., Wang, Z., Zhao, B., Luo, W., Dai, J., Lacey, S. D., Zhou, L., Shen, F., Kim, M., Swafford, L., Sengupta, L., & Hu, L., (2015). Transient rechargeable batteries triggered by cascade reactions. *Nano Lett., 15,* 4664–4671.
111. Feig, V. R., Tran, H., & Bao, Z., (2018). Biodegradable polymeric materials in degradable electronic devices. *ACS Cent. Sci., 4,* 337–348.
112. Barzic, A. I., Stoica, I., & Barzic, R. F., (2015). Microstructure implications on surface features and dielectric properties of nanoceramics embedded in polystyrene. *Rev. Roum. Chim., 60*(7/8), 809–815.
113. Barzic, R. F., Barzic, A. I., & Dumitrascu, G., (2014). Percolation network formation in poly(4-vinylpyridine)/aluminum nitride nanocomposites: Rheological, dielectric, and thermal investigations. *Polym. Compos., 35,* 1543–1552.
114. Deshmukh, K., Ahamed, M. B., Deshmukh, R. R., Pasha, K., Sadasivuni, K. K., Polu, A. R., Ponnamma, D., Al-Maadeed, M. A. A., & Chidambaram, K., (2017).

Newly developed biodegradable polymer nanocomposites of cellulose acetate and Al$_2$O$_3$ nanoparticles with enhanced dielectric performance for embedded passive applications. *J. Mater. Sci.: Mater. Electron*, *28*, 973–986.

115. Zeng, X., Deng, L., Yao, Y., Sun, R., Xu, J., & Wong, C. P., (2016). Flexible dielectric papers based on biodegradable cellulose nanofibers and carbon nanotubes for dielectric energy storage. *J. Mater. Chem. C.*, *4*, 6037–6044.

116. Hemstreet, J. M., (1982). Dielectric constant of cotton. *J. Electrost.*, *13*, 345–353.

117. Boutry, C. M., Nguyen, A., Lawal, Q. O., Chortos, A., Rondeau-Gagné, S., & Bao, Z., (2015). A sensitive and biodegradable pressure sensor array for cardiovascular monitoring. *Adv. Mater.*, *27*, 6954–6961.

118. Bao, Z., Dodabalapur, A., & Lovinger, A. J., (1996). Soluble and processable regioregular poly(3-hexylthiophene) for thin film field effect transistor applications with high mobility. *Appl. Phys. Lett.*, *69*, 4108–1210.

119. Qiao, Y., Guo, Y., Yu, C., Zhang, F., Xu, W., Liu, Y., & Zhu, D., (2012). Diketopyrrolopyrrole-containing quinoidal small molecules for high-performance, air-stable, and solution-processable n-channel organic field-effect transistors. *J. Am. Chem. Soc.*, *134*, 4084–4087.

120. Chen, H., Guo, Y., Yu, G., Zhao, Y., Zhang, J., Gao, D., Liu, H., & Liu, Y., (2012). Highly π-extended copolymers with diketopyrrolopyrrole moieties for high-performance field-effect transistors. *Adv. Mater.*, *24*, 4618–4622.

121. Pérez-Madrigal, M. M., Giannotti, M. I., Oncins, G., Franco, L., Armelin, E., Puiggalí, J., Sanz, F., del Valle, L. J., & Alemán, C., (2013). Bioactive nanomembranes of semiconductor polythiophene and thermoplastic polyurethane: Thermal, nanostructural and nanomechanical properties. *Polym. Chem.*, *4*, 568–583.

122. Pérez-Madrigal, M. M., Armelin, E., Del Valle, L. J., Estrany, F., & Alemán, C., (2012). Bioactive and electroactive response of flexible polythiophene: Polyester nanomembranes for tissue engineering. *Polym. Chem.*, *3*, 979–991.

123. Planellas, M., Pérez-Madrigal, M. M., Del Valle, L. J., Kobauri, S., Katsarava, R., Alemán, C., & Puiggalí, J., (2015). Microfibres of conducting polythiophene and biodegradable poly(ester urea) for scaffolds. *Polym. Chem.*, *6*, 925–937.

124. Lee, S., Moon, G. D., & Jeong, U., (2009). Continuous production of uniform poly(3-hexylthiophene) (P3HT) nanofibers by electrospinning and their electrical properties. *J. Mater. Chem.*, *19*, 743–748.

125. Subramanian, A., Krishnan, U. M., & Sethuraman, S., (2012). Axially aligned electrically conducting biodegradable nanofibers for neural regeneration. *J. Mater. Sci.: Mater. Med.*, *23*, 1797–1809.

126. Noriega, R., Rivnay, J., Vandewal, K., Koch, F. P., Stingelin, N., Smith, P., Toney, M. F., & Salleo, A., (2013). A general relationship between disorder, aggregation, and charge transport in conjugated polymers. *Nat. Mater.*, *12*, 1038–1044.

127. Shi, G., Rouabhia, M., Wang, Z., Dao, L. H., & Zhang, Z., (2004). A novel electrically conductive and biodegradable composite made of polypyrrole nanoparticles and polylactide. *Biomaterials*, *25*, 2477–2488.

128. Armelin, E., Gomes, A. L., Pérez-Madrigal, M. M., Puiggalí, J., Franco, L., Del Valle, L. J., Rodríguez-Galán, A., De Campos, C. J. S., Ferrer-Anglada, N., & Alemán, C., (2012). Biodegradable free-standing nanomembranes of conducting polymer:

Polyester blends as bioactive platforms for tissue engineering. *J. Mater. Chem.*, *22*, 585–594.

129. Wang, S., Guan, S., Wang, J., Liu, H., Liu, T., Ma, X., & Cui, Z., (2017). Fabrication and characterization of conductive poly(3,4-ethylenedioxythiophene) doped with hyaluronic acid/poly(L-lactic acid) composite film for biomedical application. *J. Biosci. Bioeng.*, *123*, 116–125.

130. Cui, H., Liu, Y., Deng, M., Pang, X., Zhang, P., Wang, X., Chen, X., & Wei, Y., (2012). Synthesis of biodegradable and electro active tetra aniline grafted poly(ester amide) copolymers for bone tissue engineering. *Biomacromolecules*, *13*, 2881–2889.

131. Li, M., Guo, Y., Wei, Y., MacDiarmid, A. G., & Lelkes, P. I., (2006). Electrospinning polyaniline-contained gelatin nanofibers for tissue engineering applications. *Biomaterials, 27*, 2705–2715.

132. Jeong, S. I., Jun, I. D., Choi, M. J., Nho, Y. C., Lee, Y. M., & Shin, H., (2008). Development of electroactive and elastic nanofibers that contain polyaniline and poly(L-lactide-co-e-caprolactone) for the control of cell adhesion. *Macromol. Biosci.*, *8*, 627–637.

133. Delivopoulos, E., Chew, D. J., Minev, I. R., Fawcett, J. W., & Lacour, S. P., (2012). Concurrent recordings of bladder afferents from multiple nerves using a microfabricated PDMS microchannel electrode array. *Lab Chip.*, *12*, 2540–2551.

134. Morales-Hurtado, M., Zeng, X., Gonzalez-Rodriguez, P., Ten Elshof, J. E., & Van Der Heide, E., (2015). A new water absorbable mechanical epidermal skin equivalent: The combination of hydrophobic PDMS and hydrophilic PVA hydrogel. *J. Mech. Behav. Biomed. Mater., 46*, 305–317.

135. Martino, N., Ghezzi, D., Benfenati, F., Lanzani, G., & Antognazza, M. R., (2013). Organic semiconductors for artificial vision. *J. Mater. Chem. B.*, *1*, 3768–3780.

136. Richardson-Burns, S. M., Hendricks, J. L., & Martin, D. C., (2007). Electrochemical polymerization of conducting polymers in living neural tissue. *J. Neural Eng.*, *4*, L6–L13.

137. Khodagholy, D., Doublet, T., Gurfinkel, M., Quilichini, P., Ismailova, E., Leleux, P., Herve, T., Sanaur, S., Bernard, C., & Malliaras, G. G., (2011). Highly conformable conducting polymer electrodes for *in vivo* recordings. *Adv. Mater.*, *23*, H268–H272.

138. Yang, C., Lin, W., Li, Z., Zhang, R., Wen, H., Gao, B., Chen, G., Gao, P., Yuen, M. M. F., & Wong, C. P., (2011). Water-based isotropically conductive adhesives: Towards green and low-cost flexible electronics. *Adv. Funct. Mater.*, *21*, 4582–4588.

Index

V

Valence, 37, 112, 191
Van der Waals (VDW), 69, 71, 73, 77
Velocity, 52
Vibrating sample magnetometer (VSM), 186, 189
Virtual purchase, 48, 49, 64
Volatile organic compounds (VOCs), 149

W

Wastewater
  management, 128, 138
  sampling, 34
  treatment, 29, 31, 133, 139
Water absorption test, 216, 217
Wavelength (WL), 7, 49, 76
Wiedemann-Franz (WF), 73
World
  Health Organization (WHO), 117
  War II (WW2), 52

X

X-ray
  diffraction (XRD), 7, 12, 78, 188
  photoelectron spectroscopy, 165, 166, 168, 173, 176
  powder diffraction (XRD), 1, 4, 7, 12, 13, 15, 23, 29, 34–36, 44, 78, 188
  spectrum, 189
  spectrum, 111, 112

Z

Zeolite, 74
Zeolitic imidazole framework (ZIF), 76, 78, 79
Zero
  bandgap, 73
  magnetic field, 183
Zinc oxide (ZnO), 29–32, 34–41, 43–45, 78, 79, 133
  nanoparticles, 31, 45
    characterization, 34, 35